Ship Design for Efficiency and Economy

Ship Design for Efficiency and Economy

Second edition

H. Schneekluth and V. Bertram

BUTTERWORTH
HEINEMANN

OXFORD   BOSTON   JOHANNESBURG   MELBOURNE   NEW DELHI   SINGAPORE

Butterworth-Heinemann
Linacre House, Jordan Hill, Oxford OX2 8DP
225 Wildwood Avenue, Woburn, MA 01801-2041
A division of Reed Educational and Professional Publishing Ltd

A member of the Reed Elsevier plc group

First published 1987
Second edition 1998

British Library Cataloguing in Publication Data
Schneekluth, H. (Herbert), 1921–
 Ship design for efficiency and economy.—2nd ed.
 1. Naval architecture   2. Shipbuilding
 I. Title   II. Bertram, V.
 623.8′1

ISBN 0 7506 4133 9

Library of Congress Cataloging in Publication Data
Shneekluth, H. (Herbert), 1921–
 Ship design for efficiency and economy/H. Schneekluth and
 V. Bertram. —2nd ed.
  p. cm.
 Includes bibliographical references and index.
 ISBN 0 7506 4133 9
 1. Naval architecture.   I. Bertram, V.   II. Title.
 VM770.S33   98–20741
                                                         CIP

ISBN 0 7506 4133 9

Typeset by Laser Words, Madras, India
Printed in Great Britain by Biddles Limited, Guildford and King's Lynn

Contents

Preface

This book, like its predecessors, is based on Schneekluth's lectures at the Aachen University of Technology. The book is intended to support lectures on ship design, but also to serve as a reference book for ship designers throughout their careers. The book assumes basic knowledge of line drawing and conventional design, hydrostatics and hydrodynamics. The previous edition has been modernized, reorganizing the material on weight estimation and adding a chapter on power prognosis. Some outdated material or material of secondary relevance to ship design has been omitted.

The bibliography is still predominantly German for two reasons:

- German literature is not well-known internationally and we would like to introduce some of the good work of our compatriots.
- Due to their limited availability, many German works may provide information which is new to the international community.

Many colleagues have supported this work either by supplying data, references, and programs, or by proofreading and discussing. We are in this respect grateful to Walter Abicht, Werner Blendermann, Jürgen Isensee, Frank Josten, Hans-Jörg Petershagen, Heinrich Söding, Mark Wobig (all TU Hamburg-Harburg), Norbert von der Stein (Schneekluth Hydrodynamik), Thorsten Grenz (Hapag-Lloyd, Hamburg), Uwe Hollenbach (Ship Design & Consult, Hamburg), and Gerhard Jensen (HSVA, Hamburg).

Despite all our efforts to avoid mistakes in formulas and statements, readers may still come across points that they would like to see corrected in the next edition, sometimes simply because of new developments in technology and changes to regulations. In such cases, we would appreciate readers contacting us with their suggestions.

This book is dedicated to Professor Dr.-Ing. Kurt Wendel in great admiration of his innumerable contributions to the field of ship design in Germany.

H. Schneekluth and V. Bertram

1

Main dimensions and main ratios

The main dimensions decide many of the ship's characteristics, e.g. stability, hold capacity, power requirements, and even economic efficiency. Therefore determining the main dimensions and ratios forms a particularly important phase in the overall design. The length L, width B, draught T, depth D, freeboard F, and block coefficient C_B should be determined first.

The dimensions of a ship should be co-ordinated such that the ship satisfies the design conditions. However, the ship should not be larger than necessary. The characteristics desired by the shipping company can usually be achieved with various combinations of dimensions. This choice allows an economic optimum to be obtained whilst meeting company requirements.

An iterative procedure is needed when determining the main dimensions and ratios. The following sequence is appropriate for cargo ships:

1. Estimate the weight of the loaded ship. The first approximation to the weight for cargo ships uses a typical deadweight:displacement ratio for the ship type and size.
2. Choose the length between perpendiculars using the criteria in Section 1.1.
3. Establish the block coefficient.
4. Determine the width, draught, and depth collectively.

The criteria for selecting the main dimensions are dealt with extensively in subsequent chapters. At this stage, only the principal factors influencing these dimensions will be given.

The *length* is determined as a function of displacement, speed and, if necessary, of number of days at sea per annum and other factors affecting economic efficiency.

The *block coefficient* is determined as a function of the Froude number and those factors influencing the length.

Width, *draught* and *depth* should be related such that the following requirements are satisfied:

1. Spatial requirements.
2. Stability.
3. Statutory freeboard.
4. Reserve buoyancy, if stipulated.

1

The main dimensions are often restricted by the size of locks, canals, slip-ways and bridges. The most common restriction is water depth, which always affects inland vessels and large ocean-going ships. Table 1.1 gives maximum dimensions for ships passing through certain canals.

Table 1.1 Main dimensions for ships in certain canals

Canal	L_{max} (m)	B_{max} (m)	T_{max} (m)
Panama Canal	289.5	32.30	12.04
Kiel Canal	315	40	9.5
St Lawrence Seaway	222	23	7.6
Suez Canal			18.29

1.1 The ship's length

The desired technical characteristics can be achieved with ships of greatly differing lengths. Optimization procedures as presented in Chapter 3 may assist in determining the length (and consequently all other dimensions) according to some prescribed criterion, e.g. lowest production costs, highest yield, etc. For the moment, it suffices to say that increasing the length of a conventional ship (while retaining volume and fullness) increases the hull steel weight and decreases the required power. A number of other characteristics will also be changed.

Usually, the length is determined from similar ships or from formulae and diagrams (derived from a database of similar ships). The resulting length then provides the basis for finding the other main dimensions. Such a conventional ship form may be used as a starting point for a formal optimization procedure. Before determining the length through a detailed specific economic calculation, the following available methods should be considered:

1. Formulae derived from economic efficiency calculations (Schneekluth's formula).
2. Formulae and diagrams based on the statistics of built ships.
3. Control procedures which limit, rather than determine, the length.

1. Schneekluth's formula

Based on the statistics of optimization results according to economic criteria, the 'length involving the lowest production costs' can be roughly approximated by:

$$L_{pp} = \Delta^{0.3} \cdot V^{0.3} \cdot 3.2 \cdot \frac{C_B + 0.5}{(0.145/F_n) + 0.5}$$

where:

L_{pp} = length between perpendiculars [m]
Δ = displacement [t]
V = speed (kn)
$F_n = V/\sqrt{g \cdot L}$ = Froude number

The formula is applicable for ships with $\Delta \geq 1000\,t$ and $0.16 \leq F_n \leq 0.32$.

The adopted dependence of the optimum ship's length on C_B has often been neglected in the literature, but is increasingly important for ships with small C_B. L_{pp} can be increased if one of the following conditions applies:

1. Draught and/or width are limited.
2. No bulbous bow.
3. Large ratio of underdeck volume to displacement.

Statistics from ships built in recent years show a tendency towards lower L_{pp} than given by the formula above. Ships which are optimized for yield are around 10% longer than those optimized for lowest production costs.

2. Formulae and diagrams based on statistics of built ships

1. Ship's length recommended by Ayre:

$$\frac{L}{\nabla^{1/3}} = 3.33 + 1.67\frac{V}{\sqrt{L}}$$

2. Ship's length recommended by Posdunine, corrected using statistics of the Wageningen towing tank:

$$L = C\left(\frac{V}{V+2}\right)^2 \nabla^{1/3}$$

$C = 7.25$ for freighters with trial speed of $V = 15.5\text{–}18.5\,\text{kn}$.
In both formulae, L is in m, V in kn and ∇ in m^3.
3. Völker's (1974) statistics

$$\frac{L}{\nabla^{1/3}} = 3.5 + 4.5\frac{V}{\sqrt{g\nabla^{1/3}}}$$

V in m/s. This formula applies to dry cargo ships and containerships. For reefers, the value $L/\nabla^{1/3}$ is lower by 0.5; for coasters and trawlers by 1.5.

The coefficients in these formulae may be adjusted for modern reference ships. This is customary design practice. However, it is difficult to know from these formulae, which are based on statistical data, whether the lengths determined for earlier ships were really optimum or merely appropriate or whether previous optimum lengths are still optimum as technology and economy may have changed.

Table 1.2 Length L_{pp} [m] according to Ayre, Posdunine and Schneekluth

				Schneekluth	
∇ [t]	V [kn]	*Ayre*	*Posdunine*	$C_B = 0.145/F_n$	$C_B = 1.06 - 1.68F_n$
1 000	10	55	50	51	53
1 000	13	61	54	55	59
10 000	16	124	123	117	123
10 000	21	136	130	127	136
100 000	17	239	269	236	250

In all the formulae, the length between perpendiculars is used unless stated otherwise. Moreover, all the formulae are applicable primarily to ships without bulbous bows. A bulbous bow can be considered, to a first approximation, by taking L as $L_{pp} + 75\%$ of the length of the bulb beyond the forward perpendicular, Table 1.2.

The factor 7.25 was used for the Posdunine formula. No draught limitations, which invariably occur for $\Delta \geq 100\,000\,\text{t}$, were taken into account in Schneekluth's formulae.

3. Usual checking methods

The following methods of checking the length are widely used:

1. Checking the length using external factors: the length is often restricted by the slipway, building docks, locks or harbours.
2. Checking the interference of bow and stern wave systems according to the Froude number. Unfavourable Froude numbers with mutual reinforcement between bow and stern wave systems should be avoided. Favourable Froude numbers feature odd numbers for the ratio of wave-making length L' to half-wave length $\lambda/2$ showing a hollow in the curves of the wave resistance coefficients, Table 1.3. The wave-making length L' is roughly the length of the waterline, increased slightly by the boundary layer effect.

Table 1.3 Summary of interference ratios

F_n	R_F/R_T (%)		$L':(\lambda/2)$	*Normal for ship's type*
0.19	70	Hollow	9	Medium-sized tankers
0.23	60	Hump	6	
0.25	60	Hollow	5	Dry cargo ship
0.29–0.31	50	Hump	4	Fishing vessel
0.33–0.36	40	Hollow	3	Reefer
0.40			2	
0.50	30–35	Hump	1.28	Destroyer
0.563			1	

Wave breaking complicates this simplified consideration. At Froude numbers around 0.25 usually considerable wave breaking starts, making this Froude number in reality often unfavourable despite theoretically favourable interference. The regions $0.25 < F_n < 0.27$ and $0.37 < F_n < 0.5$ should be avoided, Jensen (1994).

It is difficult to alter an unfavourable Froude number to a favourable one, but the following methods can be applied to reduce the negative interference effects:

1. Altering the length
   To move from an unfavourable to a favourable range, the ship's length would have to be varied by about half a wavelength. Normally a distortion of this kind is neither compatible with the required characteristics nor economically justifiable. The required engine output decreases when the ship is lengthened, for constant displacement and speed, Fig. 1.1. The Froude number merely gives this curve gentle humps and hollows.
2. Altering the hull form
   One way of minimizing, though not totally avoiding, unfavourable interferences is to alter the lines of the hull form design while maintaining

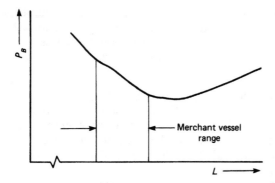

Figure 1.1 Variation of power requirements with length for constant values of displacement and speed

the specified main dimensions. With slow ships, wave reinforcement can be decreased if a prominent forward shoulder is designed one wavelength from the stem, Fig. 1.2. The shoulder can be placed at the end of the bow wave, if C_B is sufficiently small. Computer simulations can help in this procedure, see Section 2.11.

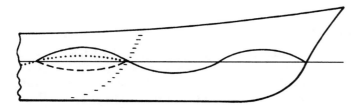

Figure 1.2 Interference of waves from bow and forward shoulder. The primary wave system, in particular the build-up at the bow, has been omitted here to simplify the presentation

3. Altering the speed
   The speed is determined largely in accordance with the ideas and wishes of the shipowner, and is thus outside the control of the designer. The optimum speed, in economic terms, can be related both to favourable and to unfavourable Froude numbers. The question of economic speed is not only associated with hydrodynamic considerations. Chapter 3 discusses the issue of optimization in more detail.

1.2 Ship's width and stability

When determining the main dimensions and coefficients, it is appropriate to keep to a sequence. After the length, the block coefficient C_B and the ship's width in relation to the draught should be determined. C_B will be discussed later in conjunction with the main ratios. The equation:

$$\nabla = L \cdot B \cdot T \cdot C_B$$

establishes the value of the product $B \cdot T$. The next step is to calculate the width as a factor in this product. When varying B at the design stage, T and D are generally varied in inverse ratio to B. Increasing B in a proposed design, while keeping the midship section area (taken up to the deck) constant, will have the following effects:

1. Increased resistance and higher power requirements: $R_T = f(B/T)$.
2. Small draught restricts the maximum propeller dimensions. This usually means lower propulsive efficiency. This does not apply if, for other reasons, the maximum propeller diameter would not be used in any case. For example, the propulsion unit may call for a high propeller speed which makes a smaller diameter essential.
3. Increased scantlings in the bottom and deck result in greater steel weight. The hull steel weight is a function of the L/D ratio.
   Items (1) to (3) cause higher production costs.
4. Greater initial stability:
   $\overline{KM}$ becomes greater, $\overline{KG}$ smaller.
5. The righting arm curve of the widened ship has steeper initial slope (resulting from the greater $\overline{GM}$), but may have decreased range.
6. Smaller draught—convenient when draught restrictions exist.

B may be restricted by building dock width or canal clearance (e.g. Panama width).

Fixing the ship's width

Where the width can be chosen arbitrarily, the width will be made just as large as the stability demands. For slender cargo ships, e.g. containerships, the resulting B/T ratios usually exceed 2.4. The L/B ratio is less significant for the stability than the B/T ratio. Navy vessels feature typical $L/B \approx 9$ and rather high centre of gravities and still exhibit good stability. For ships with restricted dimensions (particularly draught), the width required for stability is often exceeded. When choosing the width to comply with the required stability, stability conducive to good seakeeping and stability required with special loading conditions should be distinguished:

1. Good seakeeping behaviour:
   (a) Small roll amplitudes.
   (b) Small roll accelerations.
2. Special loading conditions, e.g.:
   (a) Damaged ship.
   (b) People on one side of the ship (inland passenger ships).
   (c) Lateral tow-rope pull (tugs).
   (d) Icing (important on fishing vessels).
   (e) Heavy derrick (swung outboard with cargo).
   (f) Grain cargoes.
   (g) Cargoes which may liquefy.
   (h) Deck cargoes.

Formerly a very low stability was justified by arguing that a small metacentric height $\overline{GM}$ means that the inclining moment in waves is also small. The apparent contradiction can be explained by remembering that previously the sea was considered to act laterally on the ship. In this situation, a ship with low $\overline{GM}$ will experience less motion. The danger of capsizing is also slight. Today, we know a more critical condition occurs in stern seas, especially when ship and wave speed are nearly the same. Then the transverse moment of inertia of the waterplane can be considerably reduced when the wave crest is amidships and the ship may capsize, even in the absence of previous violent motion. For this critical case of stern seas, Wendel's method is well suited (see Appendix A.1, 'German Navy Stability Review'). In this context, Wendel's experiments on a German lake in the late 1950s are interesting: Wendel tested ship models with adjustable $\overline{GM}$ in natural waves. For low $\overline{GM}$ and beam seas, the models rolled strongly, but seldomly capsized. For low $\overline{GM}$ and stern seas, the models exhibited only small motions, but capsized suddenly and unexpectedly for the observer.

Recommendations on metacentric height

Ideally, the stability should be assessed using the complete righting arm curve, but since it is impossible to calculate righting arm curves without the outline design, more easily determined $\overline{GM}$ values are given as a function of the ship type, Table 1.4. If a vessel has a $\overline{GM}$ value corresponding well to its type, it can normally be assumed (in the early design stages) that the righting arm curve will meet the requirements.

Table 1.4 Standard $\overline{GM}$—for 'outward journey', fully loaded

Ship type	$\overline{GM}$ [m]
Ocean-going passenger ship	1.5–2.2
Inland passenger ship	0.5–1.5
Tug	1.0
Cargo ship	0.8–1.0
Containership	0.3–0.6

Tankers and bulkers usually have higher stability than required due to other design considerations. Because the stability usually diminishes during design and construction, a safety margin of $\Delta \overline{GM} = 0.1$–$0.2$ m is recommended, more for passenger ships.

When specifying $\overline{GM}$, besides stating the journey stage (beginning and end) and the load condition, it is important to state whether the load condition specifications refer to grain or bale cargo. With a grain cargo, the cargo centre of gravity lies half a deck beam higher. On a normal cargo ship carrying ore, the centre of gravity is lowered by about a quarter of the hold depth. The precise value depends on the type of ore and the method of stowage.

For homogeneous cargoes, the shipowner frequently insists that stability should be such that at the end of operation no water ballast is needed. Since changeable tanks are today prohibited throughout the world, there is less tank space available for water ballast.

The $\overline{GM}$ value only gives an indication of stability characteristics as compared with other ships. A better criterion than the initial $\overline{GM}$ is the complete righting arm curve. Better still is a comparison of the righting and heeling moments. Further recommendations and regulations on stability are listed in Appendix A.1.

Ways of influencing stability

There are ways to achieve a desired level of stability, besides changing B:

(A) Intact stability

Increasing the waterplane area coefficient C_{WP}

The increase in stability when C_{WP} is increased arises because:

1. The transverse moment of inertia of the waterplane increases with a tendency towards V-form.
2. The centre of buoyancy moves upwards.

Increasing C_{WP} is normally inadvisable, since this increases resistance more than increasing width. The C_{WP} used in the preliminary design should be relatively small to ensure sufficient stability, so that adhering to a specific pre-defined C_{WP} in the lines plan is not necessary. Using a relatively small C_{WP} in the preliminary design not only creates the preconditions for good lines, but also leads to fewer difficulties in the final design of the lines.

Lowering the centre of gravity

1. The design ensures that heavy components are positioned as low as possible, so that no further advantages can be expected to result from this measure.
2. Using light metal for the superstructure can only be recommended for fast vessels, where it provides the cheapest overall solution. Light metal superstructures on cargo ships are only economically justifiable in special circumstances.
3. Installing fixed ballast is an embarrassing way of making modifications to a finished ship and, except in special cases, never deliberately planned.
4. Seawater ballast is considered acceptable if taken on to compensate for spent fuel and to improve stability during operation. No seawater ballast should be needed on the outward journey. The exception are ships with deck cargo: sometimes, in particular on containerships, seawater ballast is allowed on the outward journey. To prevent pollution, seawater ballast can only be stored in specially provided tanks. Tanks that can carry either water or oil are no longer allowed. Compared to older designs, modern ships must therefore provide more space or have better stability.

Increasing the area below the righting arm curve by increasing reserve buoyancy

1. Greater depths and fewer deckhouses usually make the vessel even lighter and cheaper. Generally speaking, however, living quarters in deckhouses are preferred to living quarters in the hull, since standardized furniture and facilities can better be accommodated in deckhouses.
2. Inclusion of superstructure and hatchways in the stability calculation. Even today, some ships, particularly those under 100 m in length, have a poop,

improving both seakeeping and stability in the inclined position, although the main reason for using a poop or a quarterdeck instead of a deckhouse is an improved freeboard. Full-width superstructures enter the water at a smaller angle of inclination than deckhouses, and have a greater effect on stability. The relevant regulations stipulate that deckhouses should not be regarded as buoyancy units. The calculations can, however, be carried out either with or (to simplify matters) without full-width superstructure. Superstructure and steel-covered watertight hatches are always included in the stability calculation when a sufficient level of stability cannot be proved without them.

3. Increasing the outward flare of framing above the constructed waterline—a flare angle of up to 40° at the bow is acceptable for ocean-going vessels.
4. Closer subdivision of the double bottom to avoid the stability-decreasing effect of the free surfaces (Fig. 1.3)
5. For ships affected by regulations concerning ice accretion, the 'upper deck purge' is particularly effective. The masts, for example, should be, as far as possible, without supports or stays.

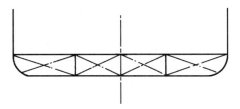

Figure 1.3 Double bottom with four-fold transverse subdivision

(B) Damaged stability

The following measures can be taken to ensure damaged stability:

1. Measures mentioned in (A) improving intact stability will also improve damaged stability.
2. Effective subdivision using transverse and longitudinal bulkheads.
3. Avoid unsymmetrical flooding as far as possible (Fig. 1.4), e.g. by cross-flooding devices.
4. The bulkhead deck should be located high enough to prevent it submerging before the permissible angle (7°–15°).

Approximate formulae for initial stability

To satisfy the variety of demands made on the stability, it is important to find at the outset a basis that enables a continuing assessment of the stability conditions at every phase of the design. In addition, approximate formulae for the initial stability are given extensive consideration.

The value $\overline{KM}$ can be expressed as a function of B/T, the value $\overline{KG}$ as a function of B/D.

A preliminary calculation of lever arm curves usually has to be omitted in the first design stage, since the conventional calculation is particularly time

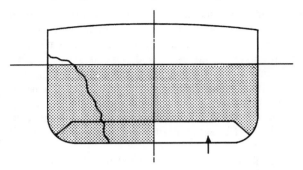

Figure 1.4 Asymmetrical flooding with symmetrical construction

consuming, and also because a fairly precise lines plan would have to be prepared for computer calculation of the cross-curves of stability. Firstly, therefore, a nominal value, dependent on the ship type and freeboard, is specified for $\overline{GM}$. This value is expected to give an acceptable lever arm curve.

The metacentric height is usually expressed as sum of three terms: $\overline{GM} = \overline{KB} + \overline{BM} - \overline{KG}$. $\overline{KG}$ will be discussed in Chapter 5, in connection with the weight calculation. Approximate formulae for $\overline{KB}$ and $\overline{BM}$ can be expressed as functions of the main dimensions, since a more precise definition of the ship's form has yet to be made at this early stage.

The main dimensions C_B, L, B, T and D are determined first. The midship section area C_M, although not fixed in the early design stages, can vary only slightly for normal ship forms and is taken as a function of C_B. Its influence on the stability is only marginal. The waterplane area coefficient C_{WP} is rarely determined before the lines design is complete and can vary greatly in magnitude depending on form (U or V sections). Its influence on the stability is considerable. Approximate values are given in Section 1.6.

Height of the centre of buoyancy above the keel

Literature on the subject has produced a series of good formulae for the value $\overline{KB}$:

Normand $\overline{KB} = T \left(\frac{5}{6} - \frac{1}{3} C_B/C_{WP} \right)$

Normand $\overline{KB} = T(0.9 - 0.36 \cdot C_M)$

Schneekluth $\overline{KB} = T(0.9 - 0.3 \cdot C_M - 0.1 \cdot C_B)$

Wobig $\overline{KB} = T(0.78 - 0.285 C_B/C_{WP})$

The accuracy of these formulae is usually better than 1% T. For the first formula, C_{WP} may be estimated from approximate formulae.

Height of metacentre above the centre of buoyancy

The approximate formulae start from the equation $\overline{BM} = I_T/\nabla$, where the transverse moment of inertia of the waterplane I_T is expressed as the moment of inertia of the circumscribing rectangle $L \cdot B^3/12$ multiplied by a reduction factor. This reduction factor is expressed as a function of C_{WP}:

$$\overline{BM} = \frac{I_T}{\nabla} = \frac{f(C_{WP}) \; L \cdot B^3}{12 \; L \cdot B \cdot T \cdot C_B} = \frac{f(C_{WP})}{12} \cdot \frac{B^2}{T \cdot C_B}$$

Approximate formulae for the reduction factor are:

Murray (trapezoidal waterplanes)	$f(C_{WP}) = 1.5 \cdot C_{WP} - 0.5$
Normand	$f(C_{WP}) = 0.096 + 0.89 \cdot C_{WP}^2$
Bauer	$f(C_{WP}) = 0.0372(2 \cdot C_{WP} + 1)^3$
N.N.	$f(C_{WP}) = 1.04 \cdot C_{WP}^2$
Dudszus and Danckwardt	$f(C_{WP}) = 0.13 \cdot C_{WP} + 0.87$
	$\cdot C_{WP}^2 \pm 0.005$

These formulae are extremely precise and generally adequate for design purposes. If unknown, C_{WP} can be estimated using approximate formulae as a function of C_B. In this way, the height of the metacentre above the centre of buoyancy $\overline{BM}$ is expressed indirectly as a function of C_B. This is always advisable when no shipyard data exist to enable preliminary calculation of C_{WP}. All formulae for $f(C_{WP})$ apply to vessels without immersed transom sterns.

Height of the metacentre above keel

$$\overline{KM} = B \left(13.61 - 45.4 \frac{C_B}{C_{WP}} + 52.17 \left(\frac{C_B}{C_{WP}} \right)^2 - 19.88 \left(\frac{C_B}{C_{WP}} \right)^3 \right)$$

This formula is applicable for $0.73 < C_B/C_{WP} < 0.95$

$$\overline{KM} = B \left(\frac{0.08}{\sqrt{C_M}} \cdot \frac{B}{T} \cdot C + \frac{0.9 - 0.3 \cdot C_M - 0.1 \cdot C_B}{B/T} \right)$$

This formula (Schneekluth) can be used without knowledge of C_{WP} assuming that C_{WP} is 'normal' corresponding to:

$$C_{WP,N} = (1 + 2C_B/\sqrt{C_M})/3$$

Then $C = 1$. If C_{WP} is better known, the formula can be made more precise by setting $C = (C_{WP,A}/C_{WP,N})^2$ where $C_{WP,A}$ is the actual and $C_{WP,N}$ the normal waterplane area coefficient.

For ships with pronounced V sections, such as trawlers or coasters, $C = 1.1–1.2$.

For a barge with a parallel-epiped form, this formula produces

for $B/T = 2$ an error $\Delta \overline{KM} = -1.6\%$, and

for $B/T = 10$ an error $\Delta \overline{KM} = +4.16\%$.

The formula assumes a 'conventional ship form' without pronounced immersed transom stern and relates to full-load draught. For partial loading, the resultant values may be too small by several per cent.

The above formula by Schneekluth is derived by combining approximate formulae for $\overline{KB}$ and $\overline{BM}$:

$$\overline{KM} = \overline{KB} + \overline{BM} = \underbrace{T \cdot (0.9 - 0.3 \cdot C_M - 0.1 \cdot C_B)}_{\text{Schneekluth}} + \underbrace{\frac{(3C_{WP} - 1)B^2}{24 \cdot C_B \cdot T}}_{\text{Murray}}$$

Substituting $C_{WP} = \frac{1}{3}(1 + 2C_B/\sqrt{C_M})$ in Murray's formula yields $\overline{BM} = 0.0834B(B/T)/\sqrt{C_M}$. Since Murray's formula can be applied exactly for trapezoidal waterplanes, (Fig. 1.5), the value must be reduced for normal waterplanes. The constant then becomes 0.08.

Figure 1.5 Comparison of ship's waterplane with a trapezium of the same area

The precision attainable using this formula is generally sufficient to determine the main dimensions. In the subsequent lines design, it is essential that $\overline{BM} = I_T/\nabla$ is checked as early as possible. The displacement ∇ is known. The transverse moment of inertia of the waterplane can be integrated numerically, e.g. using Simpson's formula.

Approximate formulae for inclined stability

At the design stage, it is often necessary to know the stability of inclined ships. The relationship

$$h = \left(\overline{BM}\frac{\tan^2 \phi}{2} + \overline{GM}\right) \sin \phi \approx \frac{1}{2}\overline{BM} \cdot \phi^3 + \overline{GM} \cdot \phi$$

('wallside formula') is correct for:

1. Wall-sided ships.
2. No deck immersion or bilge emergence.

The error due to inclined frame lines is usually smaller than the inaccuracy of the numerical integration up to 10°, provided that the deck does not immerse nor the bilge emerge. There are methods for approximating greater inclinations, but compared to the formulae for initial stability, these are more time consuming and inaccurate.

1.3 Depth, draught and freeboard

Draught

The draught T is often restricted by insufficient water depths, particularly for:

1. Supertankers.
2. Bulk carriers.
3. Banana carriers.
4. Inland vessels.

The draught must correspond to the equation $\nabla = L \cdot B \cdot T \cdot C_B$. If not restricted, it is chosen in relation to the width such that the desired degree of stability results. The advantages of large draughts are:

1. Low resistance.
2. The possibility of installing a large propeller with good clearances.

Cargo ship keels run parallel to the designed waterplane. Raked keels are encountered mostly in tugs and fishing vessels. In this case, the characteristic ratios and C_B relate to the mean draught, between perpendiculars.

Depth

The depth D is used to determine the ship's volume and the freeboard and is geometrically closely related to the draught. The depth is the cheapest dimension. A 10% increase in depth D results in an increase in hull steel weight of around 8% for $L/D = 10$ and 4% for $L/D = 14$.

The depth should also be considered in the context of longitudinal strength. If the depth is decreased, the 'flanges' (i.e. upper deck and bottom) must be strengthened to maintain the section modulus. In addition, the side-wall usually has to be strengthened to enable proper transmission of the shear forces. With the same section modulus, the same stresses are produced for constant load. But, a ship of lower depth experiences greater deflections which may damage shaftings, pipes, ceilings and other components. Consequently, the scantlings have to be increased to preserve bending rigidity when reducing depth.

Classification societies assume a restricted L/D ratio for their regulations. Germanischer Lloyd, for example, specifies a range of 10–16. However, this may be exceeded when justified by supporting calculations. Despite lower stresses, there are no further benefits for depths greater than $L/10$ as buckling may occur.

The first step when determining depth is to assume a value for D. Then this value for the depth is checked in three ways:

1. The difference between draught and depth, the 'freeboard', is 'statutory'. A 'freeboard calculation' following the regulations determines whether the assumed depth of the desired draught is permissible.
2. Then it is checked whether the depth chosen will allow both the desired underdeck volume and hold space. Section 3.4 includes approximate formulae for the underdeck volume.

3. The position of the centre of gravity, $\overline{KG}$, dependent on depth, can be verified using approximate methods or similar ships. Following this, the chosen value of the metacentric height $\overline{GM} = \overline{KM} - \overline{KG}$ can be checked.

For design purposes, an idealized depth is often adopted which is the actual depth increased by the value of the superstructure volume divided by the ship length multiplied by width.

Freeboard

The subject of freeboard has received extensive treatment in the literature, e.g. Krappinger (1964), Boie (1965), Abicht *et al.* (1974), particularly in the mid-1960s, when the freeboard regulations were re-drafted. These freeboard regulations became the object of some heavy criticism as discussed at the end of the chapter. Only the outline and the most important influencing factors of the freeboard regulations will be discussed in the following.

General comments on freeboard and some fundamental concepts

The ship needs an additional safety margin over that required for static equilibrium in calm seas to maintain buoyancy and stability while operating at sea. This safety margin is provided by the reserve of buoyancy of the hull components located above the waterline and by the closed superstructure. In addition, the freeboard is fixed and prescribed by statute. The freeboard regulations define the freeboard and specify structural requirements for its application and calculation.

The freeboard F is the height of the freeboard deck above the load line measured at the deck edge at the mid-length between the perpendiculars (Fig. 1.6). The load line is normally identical with the CWL. If there is no deck covering, the deck line is situated at the upper edge of the deck plating. If there is deck covering, the position of the deck line is raised by the thickness of the covering or a part of this.

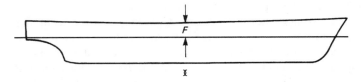

Figure 1.6 Freeboard F

The freeboard deck is usually the uppermost continuous deck, although, depending on structural requirements, requests are sometimes granted for a lower deck to be made the freeboard deck. The difference in height between the construction waterline and the uppermost continuous deck is still important in design, even if this deck is not made the freeboard deck.

Superstructures and sheer can make the freeboard in places greater than amidships. Sheer is taken into account in the freeboard regulations. The local freeboard at the forward perpendicular is particularly important (Fig. 1.7). The regulation refers to this as 'minimum bow height'. For fast ships, it is often

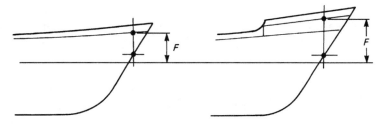

Figure 1.7 Freeboard at the forward perpendicular

advisable to make the bow higher than required in the regulations. A high bow with a small outward flare has a favourable effect on resistance in a seaway.

A 'ship with freeboard' is a ship with greater freeboard than that required by the freeboard regulation. The smaller draught resulting from the greater freeboard can be used to reduce the scantlings of the structure. For strength reasons, therefore, a 'ship with freeboard' should not be loaded to the limit of the normal permissible freeboard, but only to its own specially stipulated increased freeboard.

Effect of freeboard on ships' characteristics

The freeboard influences the following ship's characteristics:

1. Dryness of deck. A dry deck is desirable:
   (a) because walking on wet deck can be dangerous;
   (b) as a safety measure against water entering through deck openings;
   (c) to prevent violent seas destroying the superstructure.
2. Reserve buoyancy in damaged condition.
3. Intact stability (characteristics of righting arm curve).
4. Damaged stability.

A large freeboard improves stability. It is difficult to consider this factor in the design. Since for reasons of cost the necessary minimum underdeck volume should not be exceeded and the length is based on economic considerations, only a decrease in width would compensate for an increase in freeboard and depth (Fig. 1.8). However, this is rarely possible since it usually involves an undesired increase in underdeck volume. Nevertheless, this measure can be partially effected by incorporating the superstructure in the calculation of the righting arm curve and by installing full-width superstructure instead of deckhouses (Fig. 1.9).

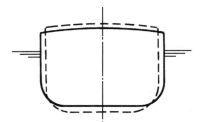

Figure 1.8 Greater freeboard at the expense of width decreases stability

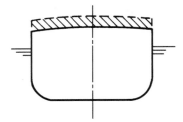

Figure 1.9 Freeboard increased by additional superstructure

Increasing depth and decreasing width would decrease both the initial stability and the righting arm curve. The stability would only be improved if the underwater form of the ship and the height of the centre of gravity remained unchanged and the freeboard were increased.

Most of the favourable characteristics which are improved by high freeboard can also be attained by other measures. However, these problems are easily solved by using adequate freeboard. Often the regulation freeboard is exceeded. Supertankers, for example, use the additional volume thus created to separate cargo and ballast compartments. Passenger ships need a higher freeboard to fulfil damage stability requirements.

The common belief that a 'good design' of a full-scantling vessel should make use of the freeboard permissible according to the freeboard calculation is not always correct. A greater than required freeboard can produce main dimensions which are cheaper than those of a ship with 'minimum freeboard'.

Freeboard and sheer

The problems associated with freeboard include the 'distribution of freeboard' along the ship's length. The sheer produces a freeboard distribution with accentuation of the ship's ends. It is here (and particularly at the forward end) that the danger of flooding caused by trimming and pitching in rough seas is most acute. This is why the freeboard regulation allows reduction of the freeboard amidships if there is greater sheer. Conversely the sheer can be decreased or entirely omitted, increasing the freeboard amidships. For constant underdeck volume, a ship without any sheer will have greater draught than a ship with normal sheer. The increase in draught depends also on the superstructure length (Fig. 1.10).

The advantages and disadvantages of a construction 'without sheer' are:

+ Better stowage of containers in holds and on deck.
+ Cheaper construction method, easier to manufacture.
+ Greater carrying capacity with constant underdeck volume.

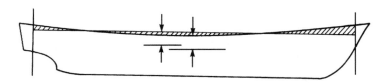

Figure 1.10 Ship with and without sheer with same underdeck volume (the differences in freeboard are exaggerated in the diagram)

- If the forecastle is not sufficiently high, reduced seakeeping ability.
- Less aesthetic in appearance.

A lack of sheer can be compensated aesthetically:

1. The 'upper edge of bulwark' line can be extended to give the appearance of sheer (Fig. 1.11).

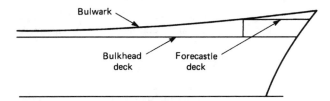

Figure 1.11 Visual sheer effect using the line of the bulwark

2. Replacement of sheer line with a suitable curved paint line or a painted fender guard (Fig. 1.12).

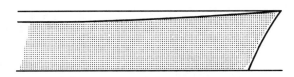

Figure 1.12 Paint line with sheer-like profile

For ships with camber of beam, care must be taken that the decks without sheer do not become too humped at the ends as a result of the deck beam curvature, i.e. the deck 'centre-line' should have no sheer and the deck edge line should be raised accordingly (Fig. 1.13). Modern cargo ships, especially those designed for container transport usually do not have camber of beam, which avoids this problem.

Figure 1.13 Forward end of deck without sheer

The International Load Line Convention of 1966

The International Load Line Convention of 1966 (ICLL 66) has been recognized by nearly every seafaring nation. The first international freeboard regulations took effect in 1904. They were modelled closely on the freeboard restrictions introduced in Great Britain in 1890 on the initiative of the British politician and social reformer Samuel Plimsoll (1824–1898). The idea of using a freeboard index line to mark this was also based on the British pattern. One particularly heavy area of responsibility was thus lifted from the shoulders of the captains. Problems associated with freeboard appeared with the emergence of steamships. Sailing vessels normally had greater freeboard to enable them to achieve the highest possible speed at greater heeling angles under sail pressure. All freeboard regulations so far have been largely based on statistically evaluated empirical data. It is difficult to demonstrate numerically to what degree the chances of the ship surviving depend on the freeboard. Hence there were widely contrasting opinions when the freeboard regulations were introduced.

The ICLL 66 is structured as follows:

Chapter I—General
All the definitions of terms and concepts associated with freeboard and the freeboard calculation, and a description of how the freeboard is marked.
Chapter II—Conditions for the assignment of freeboard
Structural requirements under which freeboard is assigned.
Chapter III—Freeboards
The freeboard tables and the regulations for correcting the basis values given by the tables. This is the most complicated and also central part of the freeboard regulations.
Chapter IV—Special regulations
For ships which are to be assigned a timber freeboard. Like Chapter II, this concerns structural requirements. These special regulations will not be discussed here.

The agreement is valid for cargo ships over 24 m in length and for non-cargo-carrying vessels, e.g. floating dredgers. An increased freeboard may be required for tugs and sailing craft. Vessels made of wood or other material or which have constructional characteristics which render an application of the regulations unreasonable or infeasible are subject to the discretion of the national authorities. The agreement states that fishing vessels need only be treated if engaged in international fish transportation or if an application for freeboard is made. Warships are not subject to the freeboard regulations.

Chapter I—General Definitions (Reg. 3)

Length—The ship's length L is the maximum of L_{pp} and 96% L_{wl}, both measured at 85% of the depth.
Perpendiculars—In the freeboard regulation, the forward perpendicular is located at the point of intersection of the waterline at 85% depth with the forward edge of the stem. The aft perpendicular is established using the rudder axis. This somewhat anomalous approach due to the forward perpendicular makes sense, since the draught (to which usually the length is related) is not available as an input value. The draught is only known after the freeboard calculation is finished.

Chapter II—Structural requirements (Regs 10–26)

The requirement for the assignment of freeboard is that the ship is sufficiently safe and has adequate strength. The requirements in detail are:

1. The national ship safety regulations must be adhered to.
2. The highest class of a recognized classification society (or the equivalent strength) must be present.
3. The particular structural requirements of the freeboard regulation must be satisfied. Particular attention should be given to: external doors, sill heights and ventilator heights, hatches and openings of every kind plus their sealing arrangements on decks and sides, e.g. engine room openings, side windows, scuppers, freeing ports and pipe outlets.

Chapter III—Freeboards

Reg. 27 of the freeboard regulations distinguishes two groups of ships:

Type A: all vessels transporting exclusively bulk liquids (tankers).
Type B: all other vessels.

Freeboard calculation procedure

The freeboard is determined as follows:

1. Determine base freeboard $F_0(L)$ according to Table 1.5.
2. Correct F_0 for $C_{B,0.85D} \neq 0.68$, $D \neq L/15$, sheer $\neq$ standard sheer, superstructures and bow height < minimum required bow height.

The corrections are:

a. Correction for ships with $24\,\text{m} < L < 100\,\text{m}$ (Reg. 29):

$$\Delta F \,[\text{mm}] = 7.5(100 - L)(0.35 - \min(E, 0.35L)/L)$$

E is the 'effective length of superstructure'. A superstructure is a decked structure on the freeboard deck, extending from side to side of the ship or with the side plating not being inboard of the shell plating more than 4% B. A raised quarterdeck is regarded as superstructure (Reg. 3(10)). Superstructures which are not enclosed have no effective length. An enclosed superstructure is a superstructure with enclosing bulkheads of efficient construction, weathertight access openings in these bulkheads of sufficient strength (Reg. 12), all other access openings with efficient weathertight means of closing. Bridge or poop can only be regarded as enclosed superstructures if access to the machinery and other working spaces is provided inside these superstructures by alternative means which are available at all times when bulkhead openings are closed. There are special regulations for trunks (Reg. 36) which are not covered here. $E = S$ for an enclosed superstructure of standard height. S is the superstructure's length within L. If the superstructure is set in from the sides of the ship, E is modified by a factor b/B_s, where b is the superstructure width and B_s the ship width, both at the middle of the superstructure length (Reg. 35). For superstructures ending in curved bulkheads, S is specially defined by Reg. 34. If the superstructure height d_v is less than standard height d_s (Table 1.5a), E is modified by a factor d_v/d_s. The effective length of a raised quarter deck (if fitted with an intact front bulkead) is its length up to a maximum of $0.6L$. Otherwise the raised quarterdeck is treated as a poop of less than standard height.

Table 1.5 Freeboard tables; intermediate lengths are determined by linear interpolation. The freeboard of ships longer than 365 m is fixed by the administration

A; tankers (Rule 28)

L (m)	F (mm)	L (m)	F (mm)	L (m)	F (mm)	L (m)	F (mm)	L (m)	F (mm)	L (m)	F (mm)
24	200	80	841	136	1736	192	2530	248	3000	304	3278
26	217	82	869	138	1770	194	2552	250	3012	306	3285
28	233	84	897	140	1803	196	2572	252	3024	308	3292
30	250	86	926	142	1837	198	2592	254	3036	310	3298
32	267	88	955	144	1870	200	2612	256	3048	312	3305
34	283	90	984	146	1903	202	2632	258	3060	314	3312
36	300	92	1014	148	1935	204	2650	260	3072	316	3318
38	316	94	1044	150	1968	206	2669	262	3084	318	3325
40	334	96	1074	152	2000	208	2687	264	3095	320	3331
42	354	98	1105	154	2032	210	2705	266	3106	322	3337
44	374	100	1135	156	2064	212	2723	268	3117	324	3342
46	396	102	1166	158	2096	214	2741	270	3128	326	3347
48	420	104	1196	160	2126	216	2758	272	3138	328	3353
50	443	106	1228	162	2155	218	2775	274	3148	330	3358
52	467	108	1260	164	2184	220	2792	276	3158	332	3363
54	490	110	1293	166	2212	222	2809	278	3167	334	3368
56	516	112	1326	168	2240	224	2825	280	3176	336	3373
58	544	114	1359	170	2268	226	2841	282	3185	338	3378
60	573	116	1392	172	2294	228	2857	284	3194	340	3382
62	600	118	1426	174	2320	230	2872	286	3202	342	3387
64	626	120	1459	176	2345	232	2888	288	3211	344	3392
66	653	122	1494	178	2369	234	2903	290	3220	346	3396
68	680	124	1528	180	2393	236	2918	292	3228	348	3401
70	706	126	1563	182	2416	238	2932	294	3237	350	3406
72	733	128	1598	184	2440	240	2946	296	3246		
74	760	130	1632	186	2463	242	2959	298	3254		
76	786	132	1667	188	2486	244	2973	300	3262		
78	814	134	1702	190	2508	246	2986	302	3270		

B (Rule 28)

L (m)	F (mm)	L (m)	F (mm)	L (m)	F (mm)	L (m)	F (mm)	L (m)	F (mm)	L (m)	F (mm)
24	200	80	887	136	2021	192	3134	248	3992	304	4676
26	217	82	923	138	2065	194	3167	250	4018	306	4695
28	233	84	960	140	2109	196	3202	252	4045	308	4714
30	250	86	996	142	2151	198	3235	254	4072	310	4736
32	267	88	1034	144	2190	200	3264	256	4098	312	4757
34	283	90	1075	146	2229	202	3296	258	4125	314	4779
36	300	92	1116	148	2271	204	3330	260	4152	316	4801
38	316	94	1154	150	2315	206	3363	262	4177	318	4823
40	334	96	1190	152	2354	208	3397	264	4201	320	4844
42	354	98	1229	154	2396	210	3430	266	4227	322	4866
44	374	100	1271	156	2440	212	3460	268	4252	324	4890
46	396	102	1315	158	2480	214	3490	270	3128	326	4909
48	420	104	1359	160	2520	216	3520	272	4302	328	4931
50	443	106	1401	162	2560	218	3554	274	4327	330	4955
52	467	108	1440	164	2600	220	3586	276	4350	332	4975
54	490	110	1479	166	2640	222	3615	278	4373	334	4995
56	516	112	1521	168	2680	224	3645	280	4397	336	5015
58	544	114	1565	170	2716	226	3675	282	4420	338	5035
60	573	116	1609	172	2754	228	3705	284	4443	340	5055
62	601	118	1651	174	2795	230	3735	286	4467	342	5075
64	629	120	1690	176	2835	232	3765	288	4490	344	5097
66	659	122	1729	178	2875	234	3795	290	4513	346	5119
68	689	124	1771	180	2919	236	3821	292	4537	348	5140
70	721	126	1815	182	2952	238	3849	294	4560	350	5160
72	754	128	1859	184	2988	240	3880	296	4583		
74	784	130	1901	186	3025	242	3906	298	4607		
76	816	132	1940	188	3062	244	3934	300	4630		
78	850	134	1979	190	3098	246	3965	302	4654		

b. Correction for $C_{B,0.85D} > 0.68$ (Reg. 30):

$$F_{\text{new}} = F_{\text{old}} \cdot (C_{B,0.85D} + 0.68)/1.36$$

The ICLL 66 generally uses the block coefficient at 0.85D, denoted here by $C_{B,0.85D}$.

c. Correction for depth D (Reg. 31):

$$\Delta F \text{ [mm]} = (D - L/15)R$$

The depth D is defined in ICLL 66 in Reg. 3(6). It is usually equal to the usual depth plus thickness of the freeboard deck stringer plate. The standard D is $L/15$. $R = L/0.48$ for $L < 120$ m and $R = 250$ for $L \geq 120$ m. For $D < L/15$ the correction is only applicable for ships with an enclosed superstructure covering at least $0.6L$ amidships, with a complete trunk, or combination of detached enclosed superstructures and trunks which extend all fore and aft. Where the height of superstructure or trunk is less than standard height, the correction is multiplied by the ratio of actual to standard height, Table 1.5a.

Table 1.5a Standard height [m] of superstructure

L [m]	Raised quarterdeck	All other superstructures
≤ 30	0.90	1.80
75	1.20	1.80
≥ 125	1.80	2.30

The standard heights at intermediate ship lengths L are obtained by linear interpolation.

d. Correction for position of deck line (Reg. 32):
   The difference (actual depth to the upper edge of the deck line minus D) is added to the freeboard. This applies to ships with rounded transitions between side and deck. Such constructions are rarely found in modern ships.

e. Correction for superstructures and trunks (Reg. 37):

$$\Delta F \text{ [mm]} = - \begin{cases} 350 + 8.3415(L - 24) & 24\,\text{m} \leq L < 85\,\text{m} \\ 860 + 5.6756(L - 85) & 85\,\text{m} \leq L < 122\,\text{m} \\ 1070 & 122\,\text{m} \leq L \end{cases}$$

This correction is multiplied by a factor depending on E (see item a) following Table 1.5b. For ships of Type B:

For $E_{\text{bridge}} < 0.2L$, linear interpolation between values of lines I and II.

For $E_{\text{forecastle}} < 0.4L$, line II applies.

For $E_{\text{forecastle}} < 0.07L$, the factor in Table 1.5b is reduced by $0.05(0.07L - f)/(0.07L)$,

where f is the effective length of the forecastle.

Table 1.5b Correction Factor for superstructures

	$E/L =$	0	0.1	0.2	0.3	0.4	0.5	0.6	0.7	0.8	0.9	1.0
Type A		0	0.07	0.14	0.21	0.31	0.41	0.52	0.63	0.753	0.877	1
Type B with	I without detached bridge	0	0.05	0.10	0.15	0.235	0.32	0.46	0.63	0.753	0.877	1
forecastle	II with detached bridge	0	0.063	0.127	0.19	0.275	0.36	0.46	0.63	0.753	0.877	1

Values for intermediate lengths E are obtained by linear interpolation.

f. Correction for sheer (Reg. 38):
The standard sheer is given by Table 1.5c. The areas under the aft and forward halves of the sheer curve are:

$$A_A = \frac{3}{48}L(y_1 + 3y_2 + 3y_3 + y_4)$$

$$A_F = \frac{3}{48}L(y_4 + 3y_5 + 3y_6 + y_7)$$

Table 1.5c Standard sheer profile [mm]

Aft Perp. (A.P.)	$y_1 = 25\left(\frac{L}{3} + 10\right)$
1/6 L from A.P.	$y_2 = 11.1\left(\frac{L}{3} + 10\right)$
1/3 L from A.P.	$y_3 = 2.8\left(\frac{L}{3} + 10\right)$
Amidships	$y_4 = 0$
Amidships	$y_4 = 0$
1/3 L from F.P.	$y_5 = 5.6\left(\frac{L}{3} + 10\right)$
1/6 L from F.P.	$y_6 = 22.2\left(\frac{L}{3} + 10\right)$
Forward Perp. (F.P.)	$y_7 = 50\left(\frac{L}{3} + 10\right)$

The 'sheer height' M is defined as the height of a rectangle of the same area: $M = (A_A + A_F)/L$. The freeboard is corrected as:

$$\Delta F = (M_{\text{standard}} - M) \cdot (0.75 - S/(2L))$$

For superstructures exceeding the standard height given in Table 1.5a, an ideal sheer profile can be used:

$$A_{A,\text{equivalent}} = \frac{1}{3}(S_A \cdot y)$$

$$A_{F,\text{equivalent}} = \frac{1}{3}(S_F \cdot y)$$

S_A is the length of the superstructure in the aft half, S_F in the fore half. y is here the difference between actual and standard height of superstructure.

This equivalent area is especially relevant to modern ships which are usually built without sheer, but with superstructures. Reg. 38 contains many more special regulations for ships with sheer which are usually not applicable to modern cargoships and not covered here.

g. Correction for minimum bow height (Reg. 39):
   The local freeboard at forward perpendicular (including design trim) must be at least:

$$F_{FP,\text{min}} \text{ [mm]} = \begin{cases} 76.16L(1 - 0.002L)/(\max(0.68, C_{B,0.85D}) + 0.68) \\ \qquad\qquad\qquad\qquad\qquad\qquad \text{for } L < 250\,\text{m} \\ 9520/(\max(0.68, C_{B,0.85D}) + 0.68) \quad \text{for } L \geq 250\,\text{m} \end{cases}$$

   If this bow height is obtained by sheer, the sheer must extend for at least 15% L abaft F.P. If the bow height is obtained by a superstructure, the superstructure must extend at least 7% L abaft F.P. For $L \leq 100\,\text{m}$, the superstructure must be enclosed.

h. The freeboard must be at least 50 mm. For ships with non-weathertight hatches the minimum freeboard is 150 mm.

The result is the Summer freeboard. This provides the basis for the construction draught and is regarded as the standard freeboard. It is the freeboard meant when using the term on its own. The other freeboard values are derived from the Summer freeboard (Reg. 40):

'Winter', 'Winter–North Atlantic', 'Tropics', 'Freshwater' and 'Freshwater Tropics'.

Criticism of the freeboard regulations

The freeboard regulations have been criticized for the following reasons:

1. For small ships, the dependence of the freeboard on ship size results in smaller freeboards not only in absolute, but also in relative terms. Seen in relation to the ship size, however, the small ship is normally subjected to higher waves than the large ship. If the freeboard is considered as giving protection against flooding, the smaller ship should surely have relatively greater freeboard than the larger ship.

   The basis freeboard for Type B ships (Fig. 1.14), ranges from less than 1% of the ship's length for small vessels up to more than 1.5% for large ships. The critics demanded freeboards of 1–2% of the length for the whole range. Advocates of the current freeboard regulation argue that:

   (a) Small vessels are engaged in coastal waters and have more chance of dodging bad weather.
   (b) The superstructures of small vessels are less exposed than those of large vessels to the danger of destruction by violent seas since sea washing on board slows the small ship down more than the large ship. Furthermore, the speeds of smaller cargo ships are usually lower than those of larger ships.
   (c) The preferential treatment given to the small ship (with respect to freeboard) is seen as a 'social measure'.

2. The freeboard regulations make the freeboard dependent on many factors such as type, size and arrangement of superstructure and sheer. The physical

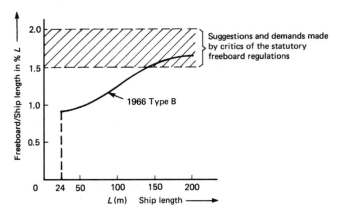

Figure 1.14 Table freeboards type B

relationships between the data entered into the calculation and their effects on ship safety are not as clear as they appear in the calculation.

3. Requiring subdivision and damage stability for larger tankers in the new freeboard regulation is generally approved, but technically it should not be part of the freeboard regulations. Furthermore, other ship types (e.g. coasters) appear to be in considerably greater danger than tankers. Meanwhile, strict subdivision rules exist for tankers in the MARPOL convention and for cargo ships over 80 m in length in the SOLAS convention.

4. The freeboard seems insufficient in many areas (particularly for small full-scantling vessels).

Unlike previous regulations, the final draft of the current freeboard regulations attempts not to impair in any way the competitive position of any ship type.

The 'minimum bow height' is seen as a positive aspect of the current freeboard regulations. Despite the shortcomings mentioned, the existing freeboard regulations undoubtedly improve safety.

New IMO freeboard regulations are being discussed and targeted to be in force by the year 2000. Alman *et al.* (1992) point out shortcomings of the ICLL 66 for unconventional ships and propose a new convention reflecting the advancements in analytical seakeeping and deck wetness prediction techniques now available. Meier and Östergaard (1996) present similar proposals for individual evaluations based on advanced seakeeping programs. They also propose simple formulae as future freeboard requirements.

Interim guidelines of the IMO for open-top containerships already stipulate model tests and calculations to determine the seakeeping characteristics. However, the interim guidelines of 1994 stipulate that under no circumstances should the freeboard and bow height assigned to an open-top containership be less than the equivalent geometrical freeboard determined from the ICLL 1966 for a ship with hatch covers.

1.4 Block coefficient and prismatic coefficient

The block coefficient C_B and the prismatic coefficient C_P can be determined using largely the same criteria. C_B, midship section area coefficient C_M

and longitudinal position of the centre of buoyancy determine the length of entrance, parallel middle body and run of the section area curve (Fig. 1.15). The shoulders become more pronounced as the parallel middle body increases. The intermediate parts (not named here) are often added to the run and the entrance.

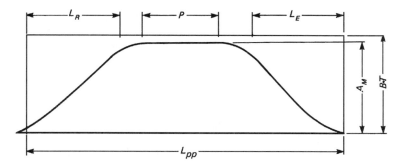

Figure 1.15 Section area curve. L_R = length of run. P = parallel middle body (range of constant sectional area and form). L_E = length of entrance

C_B considerably affects resistance. Figure 1.16 shows the resistance curve for a cargo ship with constant displacement and speed, as C_B is varied. This curve may also have humps and hollows. The usual values for C_B are far greater than the value of optimum resistance. The form factor $(1 + k)$—representing the viscous resistance including the viscous pressure resistance—generally increases with increasing C_B. Typical values for $(1 + k)$ lie around 1.13 for $C_B < 0.7$ and 1.25 for $C_B = 0.83$. In between one may interpolate linearly.

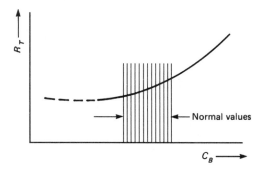

Figure 1.16 Ship's resistance as a function of the block coefficient

Shipowner requirements can be met using a wide variety of C_B values. The 'optimum' choice is treated in Chapter 3.

If C_B is decreased, B must be increased to maintain stability. These changes have opposite effects on resistance in waves, with that of C_B dominating. With lower C_B, power reduction in heavy seas becomes less necessary.

Recommendations for the choice of C_B normally draw on the statistics of built ships and are usually based on the form $C_B = K_1 - K_2 F_n$ (Alexander formula); one due to Ayre is

$$C_B = C - 1.68F_n$$

$C = 1.08$ for single-screw and $C = 1.09$ for twin-screw ships. Today, often $C = 1.06$ is used.

The results of optimization calculations provided the basis for our formulae below. These optimizations aim at 'lowest production costs' for specified deadweight and speed. The results scatter is largely dependant on other boundary conditions. In particular, dimensional restrictions and holds designed for bulky cargo increase C_B. A small ratio L/B decreases C_B:

$$C_B = \frac{0.14}{F_n} \cdot \frac{L/B + 20}{26} \qquad C_B = \frac{0.23}{F_n^{2/3}} \cdot \frac{L/B + 20}{26}$$

The formulae are valid for $0.48 \le C_B \le 0.85$ and $0.14 \le F_n \le 0.32$. However, for actual $F_n \ge 0.3$ only $F_n = 0.30$ should be inserted in the formulae.

These formulae show that in relation to the resistance, C_B and L/B mutually influence each other. A ship with relatively large C_B can still be considered to be fine for a large L/B ratio (Table 1.6). The Schneekluth formulae (lower two lines of Table 1.6) yield smaller C_B than Ayre's formulae (upper two lines), particularly for high Froude numbers. For ships with trapezoidal midship section forms, C_B should relate to the mean midship section width.

Jensen (1994) recommends for modern ship hulls C_B according to Fig. 1.17. Similarly an analysis of modern Japanese hulls gives:

$$C_B = -4.22 + 27.8 \cdot \sqrt{F_n} - 39.1 \cdot F_n + 46.6 \cdot F_n^3 \quad \text{for } 0.15 < F_n < 0.32$$

Table 1.6 C_B according to various formulae, for $L/B = 6$

Formula	Froude number F_n					
	0.14	0.17	0.20	0.25	0.30	0.32
$C_B = 1.08 - 1.68F_n$	0.85	0.79	0.74	0.66	0.58	0.54
$C_B = 1.06 - 1.68F_n$	0.83	0.77	0.72	0.64	0.56	0.52
$C_B = 0.23F_n^{-2/3}$	0.85	0.75	0.68	0.58	0.51	0.51
$C_B = 0.14/F_n$	0.85	0.82	0.72	0.56	0.48	0.48

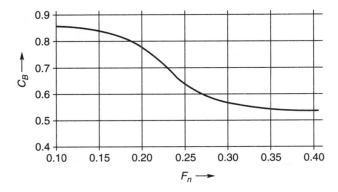

Figure 1.17 Recommended block coefficient C_B (Jensen, 1994), based on statistics

1.5 Midship section area coefficient and midship section design

The midship section area coefficient C_M is rarely known in advance by the designer. The choice is aided by the following criteria (Fig. 1.18):

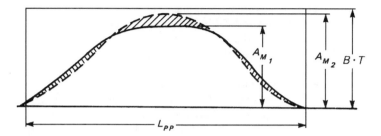

Figure 1.18 Section area curves with constant displacement and main dimensions, but different midship area coefficients

1. Effects on resistance

Increasing C_M while keeping C_B constant will usually have the following effects:

+ Increased run length—decreased separation resistance.
+ Increased entrance length—decreased wave resistance.
− Increased wetted surface area—longer flowlines, more uneven velocity distribution. Increased frictional and separation resistance.

The total influence on resistance is small, usually only a few per cent for variation within the normal limits. In designs of cargo ships where displacement and main dimensions are specified, an increase in C_M will decrease the prismatic coefficient C_P. In this case, methods for calculating resistance which use prismatic coefficient C_P will indicate a decrease in resistance, but this does not happen—at least, not to the extent shown in the calculation. The reason is that these resistance calculation methods assume a 'normal' C_M.

2. Effects on plate curvature

High C_M and the associated small bilge radii mean that the curved part of the outer shell area is smaller both in the area of the midship section and the parallel middle body. The amount of frame-bending necessary is also reduced. Both advantages are, however, limited to a small part of the ship's length. Often, the bilge radius is chosen so as to suit various plate widths.

3. Effects on container stowage

In containerships, the size and shape of the midship section are often adapted where possible to facilitate container stowage. This may be acceptable for width and depth, but is not a good policy for C_M, since this would affect only a few containers on each side of the ship.

4. Effects on roll-damping

Due to the smaller rolling resistance of the ship's body and the smaller radius of the path swept out by the bilge keel, ships with small C_M tend to experience greater rolling motions in heavy seas than those with large C_M. The simplest way to provide roll-damping is to give the bilge keel a high profile. To avoid damage, there should be a safety gap of around 1% of the ship's width between the bilge keel and the rectangle circumscribing the midship section: with rise of floor, the safety margin should be kept within the floor tangent lines. The height of the bilge keel is usually greater than 2% of the ship's width or some 30% of the bilge radius. The length of the bilge keel on full ships is approximately $L_{pp}/4$. The line of the bilge keel is determined by experimenting with models (paint-streak or wool tuft experiments) or computer simulations (CFD).

The C_M values in Table 1.7 apply only to conventional ship types. For comparison, the Taylor series has a standard $C_M = 0.925$. The C_M given in the formulae are too large for ships with small L/B. For very broad ships, keeping C_M smaller leads to a greater decrease in the wetted surface, length of flowlines and resistance.

Table 1.7 Recommendations for C_M of ships without rise of floor

for ships with	$C_B = 0.75$	$C_M = 0.987$
rise of floor	0.70	0.984
	0.65	0.980
	0.60	0.976
	0.55	0.960
for ships without		$C_M = 0.9 + 0.1 \cdot C_B$
rise of floor		$C_M = 1/(1 + (1 - C_B)^{3.5})$
		$C_M = 1.006 - 0.0056 \cdot C_B^{-3.56}$

For modern hull forms, Jensen (1994) recommends C_M according to Fig. 1.19.

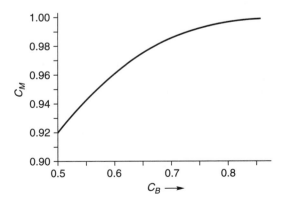

Figure 1.19 Recommended midship area coefficient C_M (Jensen, 1994)

Recommendations for bilge radius

The bilge radius R of both conventionally formed and very broad ships without rise of floor is recommended to be:

$$R = \frac{B \cdot C_K}{\left(\dfrac{L}{B} + 4\right) \cdot C_B^2}$$

$C_K = 0.5$–0.6, in extreme cases 0.4–0.7.

This formula can also be applied in a modified form to ships with rise of floor, in which case C_B should relate to the prism formed by the planes of the side-walls and the rise of floor tangents, and be inserted thus in the bilge radius formula.

$$C_B' = C_B \cdot \frac{T}{T - A/2}$$

where A is the rise of floor. The width of ships with trapezoidal midship sections is measured at half-draught (also to calculate C_B). It is usual with faster ships ($F_n > 0.4$) to make the bilge radius at least as great as the draught less rise of floor. The bilge radius of broader, shallower ships may exceed the draught.

Designing the midship section

Today, nearly all cargo ships are built with a horizontal flat bottom in the midship section area. Only for $C_M < 0.9$ is a rise of floor still found. Sometimes, particularly for small C_M, a faired floor/side-wall transition replaces the quarter circle. The new form is simpler since it incorporates a flat slipway surface and a less complicated double bottom form (Fig. 1.20). A flat bottom can be erected more cheaply on a 'panel line', and manufactured more economically.

The desired C_M is obtained by choosing a corresponding bilge radius. The bilge radius applies to ships without rise of floor and floor/side-walls transition curves:

$$R = \sqrt{2.33 \cdot (1 - C_M) \cdot B \cdot T}$$

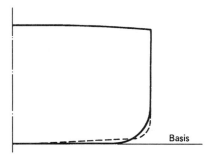

Basis

Figure 1.20 Older and more recent midship section forms

$$C_M = 1 - \frac{R^2}{2.33 \cdot B \cdot T}$$

Flared side-walls in the midships area

Cargo ships usually have vertical sides in the midship section area. Today, however, some are built with trapezoidal flared sides. The 'trapeze form' (Fig. 1.21) is more suitable than vertical sides in containerships because it improves the ratio of usable cargo hold area to overall cargo hold area. The trapeze form reduces the lateral underdeck area unusable for container stowage without necessitating a decrease in the lateral deck strips next to the hatches required for strength. Hence for a given number of containers the underdeck volume can be kept smaller than for vertical sides. When comparing with a ship with vertical sides, two cases must be distinguished in relation to resistance and power requirement:

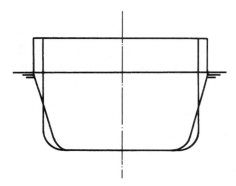

Figure 1.21 Trapezoidal midship section form

1. *Same midship section area*—In this case (at a given draught) the ship with trapezoidal midship section is broader and has, with the same prismatic coefficient C_P, a smaller C_B and a somewhat smaller wetted surface. In this comparison the ship with trapezoidal midship section usually has more favourable resistance characteristics. As ship size is increased, large containerships with trapezoidal midship sections and constant midship section areas reach the maximum Panama Canal width of $B = 32.24$ m before conventional ships with vertical sides.
2. *Same midship section dimensions*—Thus the ship with a trapezoidal midship section has a smaller midship section area, the same C_B and a higher C_P. The ship with trapezoidal midship section normally has higher resistance and power requirements.

The advantages of trapezoidal midship section can be exploited most effectively on containerships. The angle of flare of the side-walls depends on the spatial conditions and the necessary stability when empty or ballasted. At a smaller draught, the smaller second moment of area of the waterplane normally reduces the stability to such an extent that it provides a limit for the angle of flare of the side-walls. In addition, the lower ballast capacity of the double bottom further reduces the stability when ballasted. The stability when ballasted of this particular ship type must be checked early in the design. Underdeck space can be used to store fuel, and compensates for the low volume of the double bottom. A further disadvantage of the trapezoidal midship section is its exposure of the oblique sides to damage from bollards in tidal harbours. The trapezoidal midship section improves damaged stability. If the frames are flared above the load line, the second moment of area of the waterplane will increase when the ship is immersing.

There are no special resistance calculation methods for ships with trapezoidal midship sections. The resistance of these ships can be determined using the usual methods. In methods which make use of the prismatic coefficient C_P, a slight reduction (compared with normal ship forms with vertical walls and the same resistance coefficients) in the overall resistance corresponding to the reduction in the wetted surface, is produced. In methods using C_B, C_B should be based on the width at half-draught.

1.6 Waterplane area coefficient

The waterplane area coefficient C_{WP} influences resistance and stability considerably. It is geometrically closely related to the shape of cross-sections. So before making even a temporary determination of the coefficient, we should consider the sectional shapes fore and aft.

The usual procedure is to find a value for C_{WP} in the preliminary design and retain it in the lines design. There is a common tendency to use a high C_{WP} to attain a desirable degree of stability. This frequently causes unwanted distortions in lines. It is better to choose a C_{WP} at the lower limit which matches the other values, and then to design the lines independently of this. Lines which are not bound to one definite C_{WP} are not only easier to design, they generally also have lower resistance.

In the early design stages, C_{WP} is uncertain. Many approximate formulae for the stability, especially the exacter ones, contain C_{WP}. If these formulae are not to be disregarded, C_{WP} has to be estimated. The value of C_{WP} is largely a function of C_B and the sectional shape. Ships with high L/B ratio may have either U or V sections. Ships with low L/B usually have extreme V forms. Although not essential geometrically, these relationships are conventionally recognized in statistical work.

The following are some approximate formulae for C_{WP} of ships with cruiser sterns and 'cut-away cruiser sterns'. As these formulae are not applicable to vessels with submerged transom sterns, they should be tested on a 'similar ship' and the most appropriate ones adopted.

U section form, no projecting
stern form: $C_{WP} = 0.95C_P + 0.17\sqrt[3]{1 - C_P}$

Average section: $C_{WP} = (1 + 2C_B)/3$

V section form, possibly
as projecting stern form: $C_{WP} = \sqrt{C_B} - 0.025$

$C_{WP} = C_P^{2/3}$

$C_{WP} = (1 + 2C_B/\sqrt{C_M})/3$

Tanker, bulker $\qquad\qquad\qquad C_{WP} = C_B/(0.471 + 0.551 \cdot C_B)$

Table 1.8 shows examples of C_{WP} obtained by these formulae.

Table 1.8 Waterplane area coefficient values

			C_{WP}			
C_B	C_M	$(1 + 2C_B)/3$	$\begin{array}{c}0.95C_P\\+0.17\sqrt[3]{1-C_P}\end{array}$	$C_P^{2/3}$	$\sqrt{C_B} - 0.025$	$(1 + 2C_B/\sqrt{C_M})/3$
0.50	0.78	0.666	0.722	0.745	0.682	0.710
0.50	0.94	0.666	0.637	0.658	0.682	0.677
0.60	0.98	0.733	0.706	0.722	0.749	0.740
0.70	0.99	0.800	0.785	0.793	0.812	0.802
0.80	0.99	0.866	0.866	0.868	0.869	0.870

A further influence is that of the aft overhang if the values C_B and C_P relate as usual to the perpendiculars. The above formulae for a pronounced overhang can be corrected by a correction factor F:

$$F = 1 + C_P \left(0.975 \frac{L_{wl}}{L_{pp}} - 1\right)$$

The point where the line of a small stern is faired into the centre-line can be regarded as the aft endpoint of an idealized waterplane length. A length 2.5% greater than L_{pp} is 'normal'.

Where the lines have been developed from a basis ship using affine distortion, C_{WP} at the corresponding draught remains unchanged. Affine distortion applies also when length, width and draught are each multiplied by different coefficients.

For 'adding or removing' a parallel middle body, C_{WP} is easily derived from the basis design.

$$C_{WP,p} = \frac{L_v \cdot C_{WP,v} + \Delta L}{L_v + \Delta L}$$

where:

$L_v = L_{pp}$ of the basis design;
ΔL = the absolute length of the parallel middle body to be added.
The index p refers to the project ship, the index v to the basis ship.

In the affine line distortion, the $\overline{KM}$ values, obtained using C_{WP}, can be derived directly from the basis design:

$$\overline{KB}_p = \overline{KB}_v \cdot (T_p/T_v)$$
$$\overline{BM}_p = \overline{BM}_v \cdot (B_p/B_v)^3$$

1.7 The design equation

The design equation describes the displacement:

$$\Delta = \rho \cdot L \cdot B \cdot T \cdot C_B \cdot K_{\text{Appendages}}$$

ρ = density, $L \cdot B \cdot T \cdot C_B = \nabla$.

The design equation can be applied to determine the main dimensions. The initial values for the design equation can be derived from 'similar ships', formulae and diagrams and are frequently (within limits) varied arbitrarily. The desired design characteristics are greatly influenced by the ratios L/B, B/T and C_B. L/B and C_B affect the resistance, B/T the stability. The design equation is expressed in terms of these ratios. The result is an equation to determine B:

$$B = \left(\frac{\Delta \cdot B/T}{\rho \cdot C_B \cdot L/B \cdot K_{\text{Appendages}}} \right)^{1/3}$$

B is therefore the only unknown directly obtainable from the design equation. Using this, the ship's length and draught are then determined from the given ratios L/B and B/T.

Usually the resistance increases with decreasing L/B. This tendency is amplified by increasing speed. The minimum resistance for virtually all block coefficients and customary corresponding speeds is obtained for $8 < L/B < 9$. Ships with C_B higher than recommended for the Froude number should be increased in width and draught to allow a more favourable C_B.

A similar equation can be formulated for the volume up to the horizontal main deck tangent line ∇_D ('Hull volume depth') using the relationship B/D. The value B/D also provides information on the stability, as an inclination of the height of the centre of gravity above the keel ($\overline{KG}$).

$$\nabla_D = L \cdot B \cdot D \cdot C_{BD} \longrightarrow B = \left(\frac{\nabla_D \cdot B/D}{C_{BD} \cdot L/B} \right)$$

C_{BD} is the block coefficient based on the depth, or more precisely, the waterplane which is tangent to the uppermost continuous deck at its lowest point. C_{BD} will often be used in the subsequent course of the design. C_{BD} can be derived approximately from C_B based on the construction waterline, see Section 3.4.

1.8 References

ABICHT, W., ARNDT, B. and BOIE, C. (1974). Freibord. Special issue 75 Jahre Schiffbautechn. Gesellschaft, p. 187

ALMAN, P., CLEARY, W. A. and DYER, M. G. et al. (1992). The international load line convention: Crossroad to the future. Marine Technology 29/4, p. 233

BOIE, C. (1965). Kentersicherheit von Schleppern, Hansa, p. 2097

JENSEN, G. (1994). Moderne Schiffslinien. Handbuch der Werften Vol. XXII, Hansa, p. 93

KRAPPINGER, O. (1964). Freibord und Freibordvorschrift, Jahrbuch Schiffbautechn. Gesellschaft, p. 232

MEIER, H. and ÖSTERGAARD, C. (1996). Zur direkten Berechnung des Freibordes für den Schiffsentwurf. Jahrbuch Schiffbautechn. Gesellschaft, p. 254

VÖLKER, H. (1974). Entwerfen von Schiffen, Handbuch der Werften Vol. XII, Hansa, p. 17

2

Lines design

2.1 Statement of the problem

When designing cargo ships, the naval architect usually knows the main dimensions (L, B, T, C_B) and the longitudinal position of the centre of buoyancy. A minimum $\overline{KM}$ value is also frequently specified. However, for ships not affected by freeboard regulations, the designer often has relative freedom to choose C_B. Here, changes in C_B appear as variations in draught. Often the lines are considered in relation to the primary criterion of speed in calm water. The lines also influence decisively the following characteristics:

1. Added resistance in a seaway.
2. Manoeuvrability.
3. Course-keeping quality
4. Roll-damping.
5. Seakeeping ability: motion characteristics in waves, slamming effects.
6. Size of underdeck volume.

If the main data (L, B, T, C_B) are established, there remains little freedom in drawing the lines. Nevertheless, arranging the distribution of the displacement along the ship's length (i.e. the shape of the sectional area curve) and choosing the midship section area coefficient is important (Fig. 2.1). There is greater freedom in shaping the ship's ends. These points should be given particular attention:

1. Shape of the sectional area curve, prominence of shoulders.

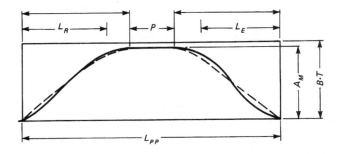

Figure 2.1 Alternative sectional area curves with the same main parameters

2. Midship section area coefficient and midship section form.
3. Bow forms, forward section forms and forward waterlines.
4. Special bow forms:
   (a) Bulbous bow.
   (b) Parabolic bow as a special form for full ships.
5. Stern forms and aft sections.

2.2 Shape of sectional area curve

Shoulder formation and a correct choice of entrance and run lengths in relation to the parallel middlebody and the position of the centre of buoyancy strongly influence the resistance coefficients. This will be dealt with in Section 2.9.

Lines of containerships

Frequently the ship's shape has to be adapted to the cargo, e.g. on ro-ro and containerships. The usual method is to fair the ship's lines around the container load plan. However, it is better to take hydrodynamically favourable ship forms and distort them linearly until all containers can be stowed as required. Minimizing the overall volume of the unoccupied spaces on containerships will not necessarily lead to greater financial savings. A bottom corner container which is too large to fit the ship's form can be accommodated by shaping the side of the containership. The shaped area can then be covered by a protrusion possessing favourable flow characteristics. This type of localized filling-out may increase resistance less than a similar procedure applied to a large area (Fig. 2.2). Placed in the forebody, the increase in resistance is negligible; in the aftbody these protrusions may generate separation.

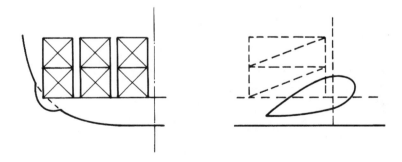

Figure 2.2 Containership with localized break in fairing

Longitudinal centres of gravity and buoyancy

The longitudinal centre of gravity can be determined from the plan of the general arrangement and ideally corresponds to the centre of buoyancy for optimum resistance. This optimum position of the centre of buoyancy is usually described in terms of a relatively broad band and as a function of C_B and the Froude number. In practice, usually the two centres of gravity and buoyancy do not coincide initially, even for the designed condition. This discrepancy usually arises when there are several load conditions, a homogeneous cargo and various draughts, e.g. for the 'open/closed shelter-decker'. The result is generally a wide range of centres of gravity for the various load conditions. Consequently, it is difficult to achieve the desired coincidence between their various longitudinal positions and that of the centre of buoyancy at designed trim, which only changes a little. The aim should be to relate the centre of gravity to the resistance-optimal centre of buoyancy, and here the whole range of recommendations can be used. Thus in developing the general design, resistance and power requirements are particularly considered. If this involves too many sacrifices with regard to volumetric design and space allocation, it may be necessary to base the centre of buoyancy on the centre of gravity instead. Often a compromise between the two extreme solutions is sought. If the centres are not co-ordinated, the ship will trim. Such trim should be kept small. With the conventional arrangement of machinery located in the aftbody, a partially loaded or empty ship will always experience stern trim, a desirable effect since it means greater propeller submergence. For a ship with machinery aft, particular attention should be paid to the trim, since the centre of the cargo is located forward of the centre of buoyancy. For light cargo there will be a tendency for stern trim. For heavy cargo the opposite is true. Figure 2.3 gives the recommended longitudinal centre of buoyancy (lcb) (taken from amidships) for ships with bulbous bows. An analysis of Japanese ships yields as typical values:

$$\text{lcb}/L = (8.80 - 38.9 \cdot F_n)/100$$

$$\text{lcb}/L = -0.135 + 0.194 \cdot C_P \text{ for tankers and bulkers}$$

Most recommendations are for resistance-optimum lcb. Power-optimum lcb are further aft.

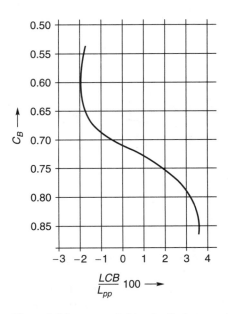

Figure 2.3 Recommended longitudinal centre of buoyancy (Jensen, 1994)

Centre of gravity of deadweight

To make the trim more independent of the cargo, the centre of the deadweight can be shifted aft by:

1. The centre of relatively heavy cargo should be moved as far aft as possible:
   (a) Foreship without sheer, forecastle only short without hold.
   (b) Collision bulkhead as far aft as possible.
   (c) High double bottom in forward hold.
   (d) Choice of propulsion system with small base area to allow forward engine room bulkhead to be located as far aft as possible.
2. Storage tanks larger than the necessary storage capacity to facilitate longitudinal transfer of fuel and fresh water for trim compensation.
3. With heavier bulk cargo not occupying all of the hold, the cargo can be stowed to locate its centre of gravity where required. This applies to such commodities as ore and crude oil.

However, heavy and light bale cargo cannot be distributed arbitrarily, neither in a longitudinal nor in a vertical direction. Normally, a ship carrying bale cargo must float on an approximately even keel with homogeneous and full loads.

2.3 Bow and forward section forms

Bows are classified as 'normal' bow, bulbous bow or special bow forms. A further distinction is made between section shapes and stem profiles. A 'normal bow' is here defined as a bow without bulb (although bulbous bows now predominate).

Stem profile

The 'normal' bow developed from the bow with vertical stem. The vertical straight stem was first used in 1840 in the United States, from where the idea quickly spread to other parts of the world. This form remained the conventional one until into the 1930s, since when it has become more raked both above and below the water. The 'dead wood' cut away reduces the resistance. The 'Maier form' used in the 1930s utilized this effect in conjunction with V sections to reduce frictional resistance (Fig. 2.4).

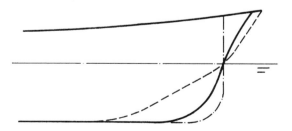

Figure 2.4 Various bulbless bow forms. —— Conventional form; - - - Maier bow of 1930s; ····· Vertical stem, in use from mid-nineteenth century to around 1930

Stems more or less raked above water offered the following advantages:

1. Water-deflecting effect.
2. Increase in reserve buoyancy.
3. Greater protection in collisions. Damage above water only more likely for both ships.
4. More attractive aesthetically (particularly when stem line is concave).

Stems with reduced rake are still used where the 'overall length' is restricted, especially on inland vessels.

Forward section shape

To characterize the section form, the letters U and V are used corresponding to the form analogy. To illustrate the various section forms, an extreme U section is compared with an extreme V section. Both must have the same sectional area below the waterline (i.e. satisfy the same sectional area curve), the same depth (up to the deck at side) and the same angle of flare at deck level (Fig. 2.5).

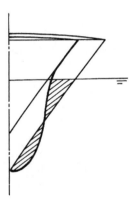

Figure 2.5 Forward U and V sections with the same underwater sectional area

Advantages and disadvantages of the V section form

+ Greater volume of topsides.
+ Greater local width in the CWL, thus greater moment of inertia of the waterplane and a higher centre of buoyancy. Both effects increase $\overline{KM}$.
+ Smaller wetted surface, lower steel weight.
+ Less curved surface, cheaper outer shell construction.
+ Better seakeeping ability due to:
  (a) Greater reserve of buoyancy.
  (b) No slamming effects.
+ Greater deck area—particularly important for the width of the forward hatch on containerships.
+ In the ballast condition at a given displacement, the wedge form increases draught and hence decreases C_B. At a smaller draught, the decreased C_B leads to a lower resistance than for U sections. Less ballast is needed to achieve the desired immersion.

— V sections in the forebody have a higher wave-making resistance with lower frictional resistance. They lead to higher overall resistance than U sections for $0.18 < F_n < 0.25$ (depending on other influencing effects of form). V sections in the forebody only have a favourable effect on resistance:
1. For normal cargo vessels, for $F_n < 0.18$ or $F_n > 0.25$.
2. For ships with $B/T > 3.5$, in a somewhat greater range.

Comparative experiments

Little has been written on the effects of the forward section form. It is a criterion rarely included in the resistance calculation. Danckwardt (1969) specifies an adjustment to the forward section depending on the position of the centre of buoyancy. The Ship Research Institute at Gothenburg investigated a ship with a U and a ship with V forward section (Institute Publication 41). The sectional area curves and main ratios were kept constant at $C_B = 0.675$, $C_M = 0.984$, $B/T = 2.4$, $L/B = 7.24$. In the 'extreme U section form' all the forward sections have vertical tangents, whereas in the 'extreme V form', the sections have comparatively straight-line forms in the forebody (Fig. 2.6). The following conclusions have been derived concerning ships without bulbous bows:

1. In the range where V sections have an optimum effect on resistance, extreme V sections should be used (Fig. 2.7).
2. In the range where U sections have an optimum effect on resistance, the advantages and disadvantages of this form must be assessed.
    (a) At points of transition between the ranges, a mean section form is used.
    (b) At the middle of the range where U sections are hydrodynamically most advantageous ($F_n = 0.23$), almost extreme U sections (Gothenburg model No. 720) are suitable.

We are not aware of any comparative experiments on U and V section forms in ships with bulbous bows, but apparently modern bulbous bows are more suited to V sections.

Forward section flare above water

Shipowners' requirements often lead to a pronounced forward section flare above water, e.g.:

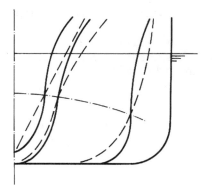

Figure 2.6 Extreme U and V section forms in the fore part of the ship (Gothenburg comparative models)

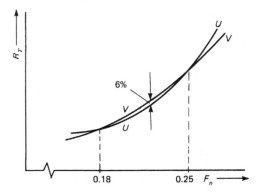

Figure 2.7 Typical resistance characteristics of U and V forms in forward sections without bulbous bow (all resistance curves intersect at two points)

1. Where there are containers on deck in the fore part of the ship.
2. Where portal crane tracks are fitted up to the forward hatch.
3. On car and train ferries where there must be a minimum entry width near to the CWL within a limited distance abaft the stem.

Increased forward section flare has these advantages and disadvantages compared to reduced flare:

+ It deflects green seas.
+ It increases the local reserve of buoyancy.
+ It reduces the pitching amplitude.
+ It increases the height of the righting arm curve.
− It can produce water spray.
− More structural material is required.
− It may lead to large pitching accelerations and impacts.

Increasing the section flare above water to raise the righting arm curve can produce good results both fore and aft. In cargo ships the forecastle sides can be flared to an angle of 40°.

Shape of the forward waterlines

The characteristic property is represented by the half-angle of entry i_E referred to the centre-line plane. i_E is related to the shape of section, sectional area curve and ship's width (Fig. 2.8). If the ship's lines are obtained by distorting an existing outline, i_E is defined automatically. Table 2.1 lists recommendations for i_E. The indicated angle has to be multiplied by the factor $7/(L/B)$. In addition, Danckwardt's resistance calculation method gives the optimum angle

Table 2.1 Recommendations for the waterline half-angle of entry based on Pophanken (1939)

C_P	0.55	0.60	0.65	0.70	0.75	0.80	0.85
i_E	8°	9°	9–10°	10–14°	21–23°	33°	37°

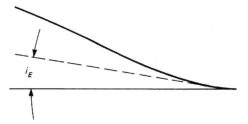

Figure 2.8 Half-angle of entry i_E of the waterline

of entry. These recommendations are primarily applicable to ships without bulbous bows and just soft guidelines.

Fore end contour of the CWL

The forward contour radius should be as small as possible in the area of the CWL. The sharpness depends on the type of construction. Round steel bars at the stem allow sufficiently small contour radii. Using sectional steel at the fore end allows a choice of sharpness. Where plates are rounded, the smallest possible radius is about 3–4 times the plate thickness. Where the stem has a round steel end bar the welded seams should be protected against ice abrasion by keeping the round steel diameter somewhat greater than that corresponding to the faired form (Fig. 2.9). In this example, the waterline plane ends short of the forward perpendicular. This shows the discrepancy that arises where the widths of the waterplane are measured to the moulded surface, but the forward perpendicular is placed at the outer edge of the stem bar. The radius at the weather deck should be relatively small, since the wave resistance rises sharply as the contour radius increases. A standard value is $R_{\mathrm{Deck}} = 0.08 \cdot B/2$ for $C_B \leq 0.72$. Downward from the waterplane, the contour radius can increase again. The transition from a round bar stem to a formed-plate stem is a costly detail of construction. A special form of bow which uses larger contour radii at the waterplane is the 'parabolic bow'.

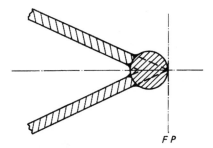

$F\,P$

Figure 2.9 Stem with round bar at the end of the CWL

Parabolic bow

Bows without sharp stems have been developed for ships with $C_B > 0.8$ and $F_n < 0.18$. They are used on tankers and bulk carriers, and also on less full vessels with high B/T ratios. These bow forms have elliptical waterlines with the minor axis of the ellipse equal to the ship's width. They are often called 'parabolic'. To improve water flow, the profile may be given a rounded form between keel and stem. These bows create a relatively large displacement in the vicinity of the perpendicular and less sharp shoulders positioned somewhat further back in comparison with alternative designs with sharp stems. Parabolic bows can also be fitted with bulbs, for which cylindrical bulb forms are usually employed. Comparative experiments using models of bulk carriers have demonstrated the superiority of parabolic bows for ships with $C_B > 0.8$ and low L/B ratios over the whole speed range investigated ($F_n = 0.11$–0.18) (Figs 2.10 and 2.11).

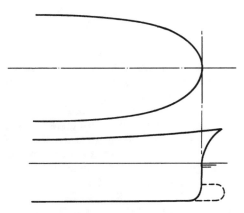

Figure 2.10 Parabolic bow—waterplane and profile

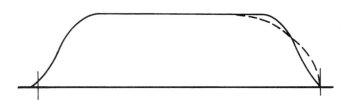

Figure 2.11 Comparison of sectional area curves of normal bow and parabolic bow

2.4 Bulbous bow

Recommended additional literature includes Hähnel and Labes (1968), Eckert and Sharma (1970), Kerlen (1971), Hoyle *et al.* (1986), and Jensen (1994).

Historical development

Today the bulbous bow is a normal part of modern seagoing cargo ships. Comparative model experiments show that a ship fitted with a bulbous bow can require far less propulsive power and have considerably better resistance characteristics than the same ship without a bulbous bow.

The bulbous bow was discovered rather than invented. Before 1900, towing tests with warships in the USA established that the ram stem projecting below the water decreased resistance. A torpedo boat model showed that an underwater torpedo discharge pipe ending in the forward stem also reduced the resistance. A bulbous bow was first used in 1912 by the US navy, based on a design by David Taylor. It was not until 1929 that the first civil ships were fitted with them. These were the passenger ships *Bremen* and *Europa* belonging to the Norddeutscher Lloyd of Bremen. A more widespread application in cargo shipping did not happen until the 1950s. The first bulb for tankers, invented by Schneekluth, was installed in 1957.

Bulbous bows are defined using the following form characteristics:

1. Shape of section.
2. Side-view.
3. Length of projection beyond perpendicular.
4. Position of axis.
5. Area ratio.
6. Transition to hull.

Some of these characteristics can be expressed by numbers.

Bulb forms

Today bulbous forms tapering sharply underneath are preferred, since these reduce slamming. The lower waterplanes also taper sharply, so that for the vessel in ballast the bulb has the same effect as a normal bow lengthened (Fig. 2.12). This avoids additional resistance and spray formation created by the partially submerged bulb. Bulbs with circular cross-sections are preferred where a simple building procedure is required and the potential danger of slamming effects can be avoided. The optimum relation of the forward section shape to the bulb is usually determined by trial and error in computer simulations, see Section 2.11 and, for example, Hoyle *et al.* (1986).

Modern bulbous forms, wedge shaped below and projecting in front of the perpendicular, are geometrically particularly well suited to V section forms.

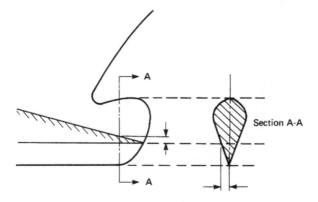

Section A-A

Figure 2.12 Modern bulb form

Cylindrical bulbs, projecting forward of the perpendicular, and Taylor non-projecting bulbs can easily be faired into U forward sections. Whether these combinations, suitable in form, lead also to minimum power requirements has yet to be discovered.

Bulbous bow projecting above CWL

It is often necessary to reduce the resistance caused by the upper side of bulbous bows which project above the CWL creating strong turbulence. The aim should be a fin effect where the upper surface of the bulb runs downwards towards the perpendicular. A bulbous bow projecting above the waterline usually has considerably greater influence on propulsion power requirements than a submerged bulb. Where a bulbous bow projects above the CWL, the authorities may stipulate that the forward perpendicular be taken as the point of intersection of the bulb contour with the CWL. Unlike well-submerged bulbs, this type of bulb form can thus increase the calculation length for freeboard and classification (Fig. 2.13). Regarding the bulb height, in applying the free-board regulations, the length is measured at 85% of the depth to the freeboard deck. Consequently, even a bulb that only approaches the CWL can still cause an increase in the calculation length of ships with low freeboard decks, e.g. shelter-deckers (Fig. 2.14).

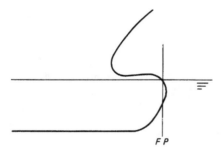

Figure 2.13 Position of forward perpendicular with high bulbous bows

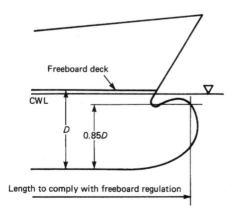

Figure 2.14 Length of freeboard calculation with low freeboard deck

Projecting length

The length projecting beyond the forward perpendicular depends on the bulb form and the Froude number. For safety reasons, the bulbous bow is never allowed to project longitudinally beyond the upper end of the stem: 20% *B* is a favourable size for the projection length. Enlarging this size improves the resistance only negligibly. Today, bulbs are rarely constructed without a projecting length. If the recess in the CWL is filled in, possibly by designing a straight stem line running from the forward edge of the bulb to the upper edge of the stem, the resistance can usually be greatly reduced. This method is hardly ever used, however.

Bulb axis

The bulb axis is not precisely defined. It should slope downwards toward the stern so as to lie in the flowlines. This criterion is also valid for the line of the maximum bulb breadth and for any concave parts which may be incorporated in the bulb. The inclination of the flowlines directly behind the stem is more pronounced in full than fine vessels. Hence on full ships, the concave part between bulb and hull should incline more steeply towards the stern.

Area ratio

The area ratio A_{BT}/A_M is the ratio of the bulb area at the forward perpendicular to the midship section area. If the bulb just reaches the forward perpendicular, or the forward edge of the bulb is situated behind the forward perpendicular the lines are faired by plotting against the curvature of the section area curve to the perpendicular (Fig. 2.15). At the design draught, the resistance of the ship with deeply submerged bulb decreases with increasing area ratio. A reduction of the area ratio (well below the resistance optimum) can, however, be advocated in the light of the following aspects:

1. Low resistance at ballast draught.
2. Avoidance of excessive slamming effects.
3. The ability to perform anchoring operations without the anchor touching the bulb.
4. Too great a width may increase the resistance of high bulbs, since these are particularly exposed to turbulence in the upper area.

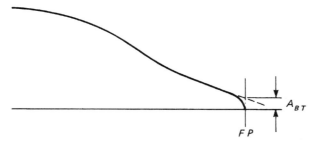

Figure 2.15 Bulb with projecting length. Theoretical bulb section area of the forward perpendicular

The effective area ratio can be further increased if the bulb is allowed to project above the CWL. Although the section above the CWL is not included in the normal evaluation of the area ratio, it increases the effective area ratio and can considerably reduce resistance, provided that the bulb is of suitable shape.

Transition

The transition from a bulbous bow to the hull can be either faired or be discontinuous (superimposed bulb). The faired-in form usually has lower resistance. The more the hollow surface lies in the flowlines, the less it increases resistance. In general, concave surfaces increase resistance less.

Bases for comparison between bulbous and normal bows

In the normal bow/bulbous bow comparison, alternative consideration and comparative model experiments usually assume a constant waterplane length between the perpendiculars.

The conventional methods to calculate the resistance of a modern vessel with bulbous bow start with a bulbless ship and then adjust to the resistance. This resistance deduction is made in only a few of the resistance calculation methods, usually insufficiently and without taking into account those bulbs with pronounced projecting forms. All resistance calculation methods can, however, include a deduction for bulbous bows using empirical values derived from any source, e.g. Kracht (1973).

The reduction in resistance can relate to the form resistance or to the overall resistance. In view of the widely differing hydrodynamic lengths of basis ships with and without bulbous bows, estimates of savings on power due to the bulbous bow are considerably less reliable than for earlier bulbous forms, which only extended to the forward perpendicular. The bulb may reduce resistance in the range $0.17 \leq F_n \leq 0.7$. Earlier non-projecting bulbs decreased resistance at best by some 6%. Modern bulbs decrease resistance often by more than 20%. Whereas above $F_n = 0.23$ the main effect of the bulb is to shift the bow wave forward, the voluminous bulbs and relatively short wavelengths of slower vessels may also cause displacement to shift forward from the area of the forward shoulder. In this way, the bulb displacement can be used to position the forward shoulder further aft, so that the entrance length approximates to the wavelength (Fig. 2.16). Another way to decrease resistance is to reduce trim at the stern.

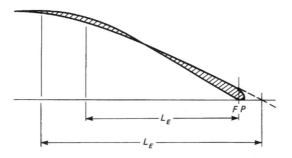

Figure 2.16 Possible increase in effective entrance length with bulbous bow

Effects of bulbous bows on ships' characteristics

The effects of a bulbous bow can extend to several areas of the ship's design, construction, manufacture and operation, e.g.:

1. Effective drag (total resistance) and characteristics at various draughts.
2. Resistance in a seaway.
3. Seakeeping characteristics.
4. Propulsion characteristics.
5. Course-keeping ability and manoeuvrability.
6. Bow-thruster:
   (a) Possibilities for installation.
   (b) Efficiency.
   (c) Additional resistance.
7. Trim.
8. Construction, manufacture and building costs of bow section.
9. Freeboard.
10. Anchor-handling apparatus and operation with respect to danger of anchor striking bulbous bow.
11. Accommodation of sounding devices on fishing and research vessels.
12. Observing length restrictions due to docks and locks.
13. Ice operation.

Of these characteristics, the following have been selected for closer examination:

1. Ice operation with bulbous bow

A certain ice-breaking capability can be achieved if the position of the upper side of the bulb enables it to raise an ice sheet. For operation in medium-thick ice, the bulbous bow has greater advantages than conventional, and even ice-breaking, bows because it turns the broken lumps so that their wet sides slide along the hull, thus causing less wear on the outer shell and less resistance. The maximum thickness which a bulbous bow can break is less than for special ice-breaking bow forms.

2. Seakeeping characteristics with bulbous bow

Three characteristics are of interest here:

1. Damping of pitching motion.
   Generally speaking, bulbous bows increase pitch motion damping, especially when designed for the purpose. The damping is particularly pronounced in the area of resonance when the wavelength roughly corresponds to the ship's length. There is even some damping for shorter wavelengths. For wavelengths exceeding 1.3–1.5 ship's lengths, ships with bulbous bows will experience an increase in pitch amplitude. However, the pitch amplitude in this range is small in relation to the wave height.
2. The ability to operate without reduction of power even in heavier seas.
   Sharp-keeled bulbs can withstand slamming effects in more severe seas than normal bulbs. Where the bulbous bow has a flat upper surface, water striking the bow may cause pounding.
3. The increased power requirements in waves.

Bulbous bows increase the added resistance due to waves, despite the smoother operation in heavy seas. This is analogous to the effect of the bilge keel. The energy of damping has to be taken from the propulsive power. For wavelengths shorter than 0.9L the pitching frequency of the ship is subcritical. Then the bulb may reduce the added resistance.

3. Power requirements with bulbous bow

The change in power requirement with the bulbous bow as opposed to the 'normal' bow can be attributed to the following:

1. Change in the pressure drag due to the displacing effect of the bulb and the fin effect.
   The bulb has an upper part which acts like a fin (Fig. 2.17). This fin-action is used by the 'stream-flow bulb' to give the sternward flow a downward component, thus diminishing the bow wave. Where the upper side of the bulb rises towards the stem, however, the fin effect decreases this resistance advantage. Since a fin effect can hardly be avoided, care should be taken that the effect works in the right direction. Surprisingly little use is made of this resistance reduction method.

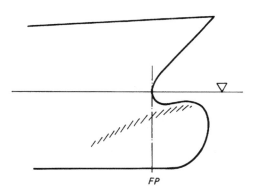

Figure 2.17 Fin bulb

2. Change in wave breaking resistance.
   With or without bulb, spray can form at the bow. By shaping the bow suitably (e.g. with sharply tapering waterlines and steep sections), spray can be reduced or completely eliminated.
3. Increase in frictional resistance.
   The increased area of the wetted surface increases the frictional resistance. At low speeds, this increase is usually greater than the reduction in resistance caused by other factors.
4. Change in energy of the vortices originating at the bow.
   A vortex is created because the lateral acceleration of the water in the CWL area of the forebody is greater than it is below. The separation of vortices is sometimes seen at the bilge in the area of the forward shoulder. The bulbous bow can be used to change these vortices. This may reduce energy losses due to these vortices and affect also the degree of energy recovery by the propeller (Hoekstra, 1975).

5. Change in propulsion efficiency influenced by:
   (a) Thrust loading coefficient.
   (b) Uniformity of flow velocity.
   In comparative experiments on models with and without bulbous bows, those with bulbous bows show usually better propulsion characteristics. The obvious explanation, i.e. that because the resistance is lower, a lower thrust coefficient is also effective, which leads to higher propeller efficiency in cargoships, is correct but not sufficient. Even at speeds where the resistances are equal and the propeller thrust loading coefficients roughly similar, there is usually an improvement of several per cent in the bulbous bow alternative (Fig. 2.18). Kracht (1973) provides one explanation of why the bulb improves propulsion efficiency. In comparative experiments, he determined a greater effective wake in ships with bulbous bows. Tzabiras (1997) comes to the same conclusion in numerical simulations for tanker hull forms.

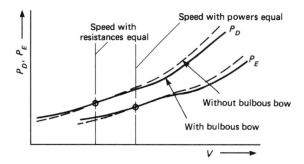

Figure 2.18 Resistance comparison (ship with and without bulbous bow)

The power savings by a bulbous bow may, depending on the shape of the bulb, increase or decrease with a reduction in draught. The lower sections of modern bulbous bows often taper sharply. The advantage of these bulbous bows is particularly noticeable for the ship in ballast.

Criteria for the practical application of bulbous bows

Writers on the subject deal with the bulbous bow almost exclusively from the hydrodynamic point of view, ignoring overall economic considerations. The power savings of a bulbous bow should be considered in conjunction with the variability of the draught and sea conditions. The capital expenditure should also be taken into account. The total costs would then be compared with those for an equivalent ship without bulbous bow. Selection methods such as these do not yet exist. The following approach can be used in a more detailed study of the appropriate areas of application of bulbous bows.

Most of the procedures used to determine a ship's resistance are based on forms without a bulbous bow. Some allow for the old type of bulbous bow where the bulb was well submerged and did not project beyond the perpendicular. A comparison between ships with and without bulbous bow usually assumes waterlines of equal length, as is the case when considering alternatives or conducting comparative experiments with models. The usual method of calculating the resistance of a modern ship with a bulbous bow is to take a ship without a bulb and then make a correction to the resistance. Some methods include this correction, others rely on collecting external data to perform the correction. The change in a ship's resistance caused by the bulbous bow depends both on the form and size of the bulb and on the form and speed of the ship.

One way of ascertaining the effect of modern bulbs on resistance is to use a 'power-equivalent length' in the calculation instead of L_{pp} or L_{wl}. The 'equivalent length' is the length of a bulbless ship of the same displacement with the same smooth-water resistance as the ship with a bulb. The equivalent length is a function of bulb form, bulb size, Froude number, and block coefficient. If bulb forms are assumed to be particularly good and the bulb is of normal size to ensure compatibility with the other desired characteristics, the resulting equivalent length will range from being only slightly greater than L_{pp} for small Froude numbers to L_{pp} plus three bulb lengths for $F_n > 0.3$. The equivalent length of conventional cargoships with Froude numbers below $F_n \approx 0.26$ is shorter than the hydrodynamic length, i.e. shorter than L_{pp} increased by the projecting part of the bulb (Fig. 2.19).

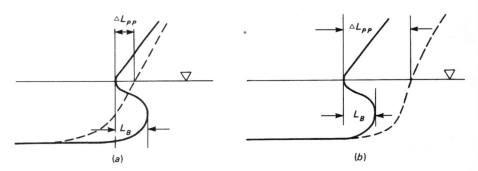

Figure 2.19 Power-equivalent bow forms. (a) Froude range $F_n = 0.22$–0.25: (b) Froude range $F_n = 0.30$–0.33

For $0.29 < F_n < 0.32$, lengthening the CWL of smaller ships reduces the power more than a bulbous bow corresponding to the CWL lengthening. However, a bulbous bow installed on ships with $F_n > 0.26$ reduces power more than lengthening the waterplane by the projecting length of the bulb. Figure 2.20 shows how far a normal bow (without bulb) must be lengthened by ΔL_{pp} to save the same amount of power as a bulbous bow, where L_B is the length of the bulb which projects beyond the perpendicular and ΔL_{pp} is the power-equivalent lengthening of the normal form. On the upper boundary of the shaded area are located ships which have a high or too high C_B in relation to F_n and vice versa. For $F_n < 0.24$ the equivalent increase in length is always less than the length of the bulbous bow. For $F_n > 0.3$, the bulb effect may not be achieved by lengthening. Thus determining an equivalent length is useful when deciding whether or not a bulbous bow is sensible.

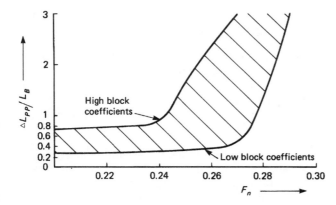

Figure 2.20 Power-equivalent lengthening of normal bow to obtain bulbous effect as a function of the Froude number

Steel-equivalent length

This is the length of a ship without bulbous bow which produces the same hull steel weight as the ship of equal displacement with bulbous bow (Fig. 2.21).

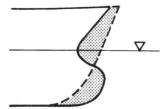

Figure 2.21 Steel weight equivalent bow forms

Conclusions of the equivalent lengths study

The problem of finding an 'optimal' length can be simplified by taking only the main factors into account and comparing only a few of the possible alternatives. Considerations can be restricted by making only the normal contractual conditions the basis of these considerations. Seakeeping and partial loading can then be disregarded for the time being. The normal procedure in this case is to compare a ship without bulb with the same ship with bulb, and to determine the decrease in propulsion power. More appropriate are comparisons of cost-equivalent or power-equivalent forms. Here, the following distinctions are made:

1. The ship is designed as a full-decker, so attention must be paid to the freeboard.
2. The ship is not governed by freeboard considerations within the range implied by a small increase in length.

The freeboard is a limiting factor for full-deckers. The vessel cannot simply be built with a lengthened normal forebody instead of a bulbous bow without increasing the freeboard and reducing the draught. Other kinds of compensation designed to maintain the carrying capacity, e.g. greater width and greater depth for the ship without bulbous bow, are disregarded for the time being. If the freeboard is not limiting, there is greater freedom in optimization:

1. A proposed bulbous bow ship can be compared with a ship with normal bow and the normal bow lengthened until power equivalence is achieved. This shows immediately which alternative is more costly in terms of steel. All other cost components remain unchanged.
2. The propulsion power can be compared to that for a normal bow with the equivalent amount of steel.

In both cases the differences in production costs and in the ship's characteristics can be estimated with reasonable precision, thus providing a basis for choice. Throughout, only ships with bulbs projecting forward of the perpendicular and ships with normal bulbless bows have been compared. If a comparative study produces an equivalence of production costs or power, then the ship without bulbs will suffer smaller operational losses in a seaway (depending on the type of bulb used in the comparative design) and possess better partial loading characteristics. A more extensive study would also examine non-projecting bulbous bows. The savings in power resulting from these can be estimated more precisely, and are within a narrower range.

2.5 Stern forms

The following criteria govern the choice of stern form:

1. Low resistance.
2. High propulsion efficiency.
   (a) Uniform inflow of water to propeller.
   (b) Good relationship of thrust deduction to wake (hull efficiency η_H).
3. Avoiding vibrations.

Development of stern for cargo ships

In discussing stern forms, a distinction must be made between the form characteristics of the topside and those of the underwater part of the vessel. The topside of the cargo ship has developed in the following stages (Fig. 2.22):

1. The merchant or elliptical stern.
2. The cruiser stern.
3. The transom stern.

In addition there are numerous special forms.

The elliptical stern

Before about 1930, the 'merchant stern', also known as elliptical or 'counter' stern, was the conventional form for cargo ships. Viewed from above, the deck line and the knuckle line were roughly elliptical in shape. The length between the perpendiculars of the merchant stern is identical with the length of the waterplane. The stern is still immediately vertical above the CWL, then flares sharply outwards and is knuckled close to the upper deck. A somewhat modified form of the merchant stern is the 'tug stern', where the flaring at the upper

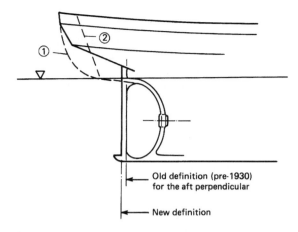

Figure 2.22 Stern contours on cargo ships. Elliptical, cruiser stern (1) and transom stern (2)

part of the stern is even more pronounced (Fig. 2.23). The knuckle occurs at the height of the upper deck. The bulwark above it inclines inwards. This form was still used on tugs and harbour motor launches after World War II.

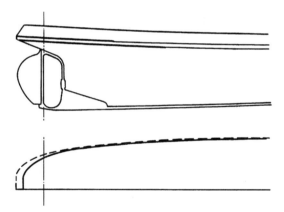

Figure 2.23 Tug stern

Cruiser stern

The cruiser stern emerged in the latter half of the nineteenth century in warships, and was initially designed only to lower the steering gear below the armour deck, located at approximately the height of the CWL. The knuckle above the CWL disappeared. The cruiser stern had better resistance character- istics than the merchant stern and consequently found widespread application on cargo ships. The length of the waterplane with a cruiser stern is greater than L_{pp}. The transition from merchant stern to cruiser stern on cargo ships took place between the world wars. The counter, situated lower than on the merchant stern, can be used to reduce resistance chiefly on twin-screw and single-screw vessels with small propeller diameters.

Transom stern

The term transom stern can be understood both as a further development of the cruiser stern and as an independent development of a stern for fast ships. The further development of the cruiser stern is effected by 'cutting off' its aft-most portion. The flat stern then begins at approximately the height of the CWL. This form was introduced merely to simplify construction. The transom stern for fast ships should aim at reducing resistance through:

1. The effect of virtual lengthening of the ship.
2. The possibility this creates of countering stern trim.

The trim can be influenced most effectively by using stern wedges (Fig. 2.24). The stern wedge gives the flow separation a downward component, thereby decreasing the height of the wave forming behind the ship and diminishing the loss of energy. The stern wedge can be faired into the stern form. As a result of the stern wedge influencing the trim, the bow is pressed deeper into the water at high speeds, and this may have a negative effect on seakeeping ability.

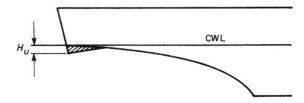

Figure 2.24 Transom stern with stern wedge

Recommendations for transom stern design

$F_n < 0.3$ Stern above CWL. Some stern submergence during operation.

$F_n \approx 0.3$ Small stern—only slightly below CWL.

$F_n \approx 0.5$ Deeper submerging stern with average wedge.

Submergence $t = 10–15\%T$.

$F_n > 0.5$ Deep submerging stern with wedge having approximately width of ship.

Submergence $t = 15–20\%T$.

Further with regard to the deeply submerged square stern:

1. The edges must be sharp. The flow should separate cleanly.
2. Ideally, the stability rather than the width should be kept constant when optimizing the stern. However, this does not happen in practice. The ship can be made narrower with a transom stern than without one.

3. The stern, and in particular its underside, influences the propulsion efficiency. There is less turbulence in the area between propeller and outer
   shell above the propeller.
4. Slamming rarely occurs. In operation, the flowlines largely follow the ship's
   form.
5. During slow operation, strong vortices form behind the transom, causing it
   to become wet. The resistance in slow-speed operation is noticeably higher
   than that of the same ship with cruiser stern.
6. The centre of pitching is situated at roughly one-quarter of the ship's length
   from aft as opposed to one-third of the length from aft on normal vessels.
   The forward section gets wetter in heavy seas.
7. The deck on transom stern ships can easily get wet during reversing operations and in a heavy sea. The water is 'dammed up'. Flare and knuckle
   deflect the water better during astern operations avoiding deck flooding
   (Fig. 2.25).

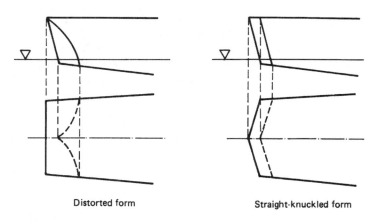

Distorted form   Straight-knuckled form

Figure 2.25 Transom stern with flared profiles

The reduction in power compared with the cruiser increases with the Froude
number. Order of magnitude: approx. 10% at $F_n = 0.5$. This reduction in
power is less due to reducing the resistance than to improving the propulsive
efficiency.

Advice on designing the stern underwater form

Attention should be paid to the following:

1. Minimizing flow separation.
2. Minimizing the suction effect of the propeller.
3. Sufficient propeller clearance.

Separation at the stern

Separation at the stern is a function of ship form and propeller influence. The
suction effect of the single-screw propeller causes the flowlines to converge.

This diminishes or even prevents separation. The effect of the propeller on twin-screw ships leads to separation. Separation is influenced by the radius of curvature of the outer shell in the direction of flow, and by the inclination of flow relative to the ship's forward motion. To limit separation, sharp shoulders at the stern and lines exceeding a critical angle of flow relative to the direction of motion should be avoided. If the flow follows the waterlines rather than the buttocks, a diagonal angle or a clearly definable waterline angle is usually the criterion instead of the direction of flow, which is still unknown at the design stage. The critical separation angles between waterline and longitudinal axis for cruiser sterns and similar forms are:

$i_R = 20°$ according to Baker—above this, separation is virtually inevitable.

$i_R = 15°$ according to Kempf—separation beginning.

An angle of less than 20° to the longitudinal axis is also desirable for diagonal lines. Adherence to these two angles is often impossible, particularly for full hull forms. Most critical is the lower area of the counter, the area between the counter and the propeller post (Fig. 2.26). In areas where the flow mainly follows the buttocks, no separation will occur, regardless of the waterline angle. This happens, for example, below a flat, transom stern and in the lower area of the stern bulb. If a plane tangential to the ship's form is assumed, the angle between longitudinal axis of the ship and this tangential plane should be as small as possible.

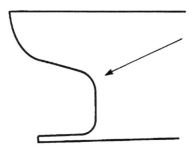

Figure 2.26 Position of greatest waterline angle

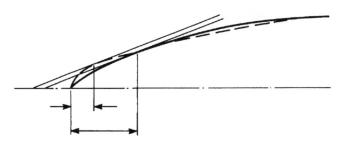

Figure 2.27 Separation zone with stern waterlines, above the propeller

The stern waterlines above the propeller should be straight, and hollows avoided, to keep waterline angles as small as possible. Where adherence to the critical waterline angle is impossible, greatly exceeding the angle over a short distance is usually preferred to marginally exceeding it over a longer distance. This restricts the unavoidable separation zone (Fig. 2.27) to a small area.

The waterline endings between counter and propeller shaft should be kept as sharp as possible (Fig. 2.28). The outer shell should run straight, or at most be lightly curved, into the stern. This has the following advantages:

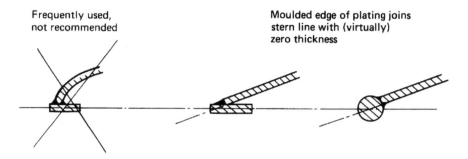

Frequently used, not recommended

Moulded edge of plating joins stern line with (virtually) zero thickness

Figure 2.28 Plating—stern post connections

1. Reduced power requirements. Reduced resistance and thrust deduction fraction.
2. Quieter propeller operation.

Methods of reducing waterline angles

Single-screw ships

If the conventional rudder arrangement is dispensed with, the inflow angle of the waterlines in the stern post area of single-screw ships can be effectively reduced by positioning the propeller post further aft. The following arrangements may be advantageous here:

1. Nozzle rudder with operating shaft passing through the plane of the tips of the propeller blades. The nozzle rudder requires more space vertically than the nozzle built into the hull, since the propeller diameter to be accommodated is smaller. The gap between propeller blade tip and nozzle interior must also be greater in nozzle rudders. For these two reasons, propulsion efficiency is not as high as with fixed nozzles.
2. Rudder propeller—and Z propulsion.

Centre-line rudder with twin-screws

Where twin-screw ships are fitted with a central rudder, it is advisable to make the rudder thicker than normal. In this way, the rudder has a hull-lengthening effect on the forward resistance of the ship. This results in a lower resistance and higher displacement with steering characteristics virtually unchanged. The ratio of rudder thickness to rudder length can be kept greater than normal (Fig. 2.29).

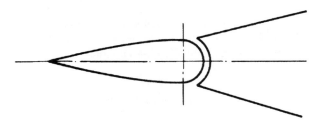

Figure 2.29 Centre-line rudder on twin-screw ships

Propeller suction effect

The lines in the area where the flow enters the propeller must be designed such that the suction remains small. Here the propeller regains some of the energy lost through separation. The following integral should be as small as possible for the suction effect (Fig. 2.30):

$$\int \frac{\sin \alpha}{a^x} \, \mathrm{d}S$$

where:

  $\mathrm{d}S$ is the surface element of the outer shell near the propeller,
  α is the angle of the surface element to the longitudinal axis of the ship,
  a is the distance of the surface element from the propeller, $x \approx 2$.

Hence it is important to keep the waterlines directly forward of the propeller as fine as possible. The waterlines forward of the propeller can be given light hollows, even if this causes a somewhat greater maximum waterline angle than straight lines. Another way of minimizing the suction is to increase the clearance between the propeller post and the leading edge of the blade.

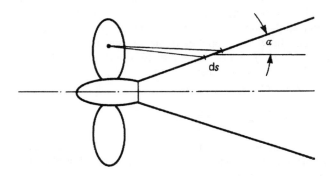

Figure 2.30 Effect of propeller suction on shell element

Wake distribution as a function of ship's form

A non-uniform inflow reduces propulsion efficiency. In predictive calculations, the propeller efficiency η_0 is derived by systematic investigations which assume an axial regular inflow. The decrease in propulsion efficiency caused by the irregular direction and velocity of the inflow is determined using the 'relative rotative efficiency' η_R in conjunction with other influencing factors. As well as diminishing propeller efficiency, an irregular wake can also cause vibrations. Particular importance is attached to the uniformity of flow at a constant radius at various angles of rotation of the propeller blade. Unlike tangential variations, radial variations in inflow velocity can be accommodated by adjusting the propeller pitch. The ship's form, especially in the area immediately forward of the propeller, considerably influences the wake distribution. Particularly significant here are the stern sections and the horizontal clearance between the leading edge of the propeller and the propeller post. See Holden *et al.* (1980) for further details on estimating the influence of the stern form on the wake.

In twin-screw ships, apart from the stern form, there are a number of other influential factors:

1. Shaft position (convergent–divergent horizontal-inclined).
2. Shaft mounting (propeller brackets, shaft bossings, Grim-type shafts).
3. Distance of propellers from ship centre-line.
4. Size of clearance.

Stern sections

The following underwater sections of cruiser and merchant sterns are distinguished (Fig. 2.31):

1. V-section.
2. U-section.
3. Bulbous stern.

On single-screw vessels, each stern section affects resistance and propulsion efficiency differently. The V section has the lowest resistance, irrespective of Froude number. The U section has a higher and the bulbous form (of conventional type) the highest resistance. Very good stern bulb forms achieve the same resistance as U-shaped stern section. On the other hand, the V section has the most non-uniform and the bulbous form the most uniform wake distribution,

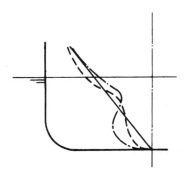

Figure 2.31 Stern sections

thus higher propulsion efficiency and less vibration caused by the propeller. This may reduce required power by up to several per cent. Therefore single-screw ships are given U or bulbous sections rather than the V form. The disadvantage of the bulbous stern is the high production cost. The stern form of twin-screw ships has little effect on propulsive efficiency and vibration. Hence the V form, with its better resistance characteristics, is preferred on twin-screw ships.

Bulbous sterns, installed primarily to minimize propeller-induced vibrations, are of particular interest today. The increased propulsive efficiency resulting from a more uniform inflow is offset by an increased resistance. Depending on the position and shape of the bulbous section, the ship may require more or less power than a ship with U section.

The bulbous stern was applied practically in 1958 by L. Nitzki who designed a bulb which allowed the installations of a normal (as opposed to one adapted to the shape of the bulb) propeller. To increase wake uniformity, he gave the end of the bulb a bulged lower section which increased the power requirement.

A later development is the 'simplified' bulbous stern (Fig. 2.32). Its underside has a conical developable form. The axis of the cone inclines downwards towards the stern and ends below the propeller shaft. The waterplanes below the cone tip end as conic sections of relatively large radius. Despite this, the angle to the ship's longitudinal axis of the tangent plane on the bulb underside is only small. With this bulb form, a greater proportion of the slower boundary-layer flow is conducted to the lower half of the propeller. The waterplanes above the bulb end taper sharply into the propeller post. The angle of run of the waterplanes at the counter can be decreased by chamfering the section between bulb and hull (Wurr, 1979). This bulbous stern has low power requirement, regular wake and economical construction.

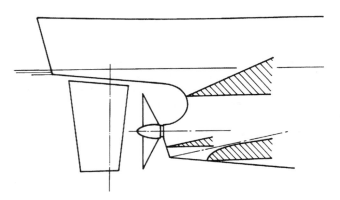

Figure 2.32 'Simplified' bulbous stern

2.6 Conventional propeller arrangement

Ship propellers are usually fitted at the stern. Bow propellers are less effective if the outflow impinges on the hull. This exposes the hull to higher frictional resistance. Bow propellers are used only on:

1. Icebreakers to break the ice by the negative pressure field in front of the propeller.
2. Double-ended ferries, which change direction frequently.
3. Inland vessels, where they act as rudder propellers. In forward operation, the forward propeller jets are directed obliquely so that they clear the hull.

Propellers are usually placed so that the gap between the upper blade tip and the waterplane is roughly half the propeller diameter. This ensures that there will still be sufficient propeller submergence at ballast draught with aft trim.

On single-screw vessels, the shaft between the aft peak bulkhead and the outer shell aperture passes through the stern tube, at the aft end of which is the stern tube bearing, a seawater-lubricated journal bearing. The inside of the inner end of the stern tube is sealed by a gland. Oil-lubricated stern tube bearings sealed off from seawater and the ship's interior are also currently in use. On twin-screw ships, the space between outer shell and propeller is so large that the shaft requires at least one more mounting. The shaft can be mounted in one of three ways—or a combination of them:

1. Shaft struts.
2. Shaft bossings with local bulging of the hull.
3. Grim-type shafts (elastic tubes carrying the shafts with a journal bearing at the aft end).

2.7 Problems of design in broad, shallow-draught ships

Ships with high B/T ratios have two problems:

1. The propeller slipstream area is small in relation to the midship section area. This reduces propulsion efficiency.
2. The waterline entrance angles increase in comparison with other ships with the same fineness $L/\nabla^{1/3}$. This leads to relatively high resistance.

Ways of increasing slipstream area

1. *Multi-screw propulsion* can increase propulsion efficiency. However, it reduces hull efficiency, increases resistance and costs more to buy and maintain.
2. *Tunnels* to accommodate a greater propeller diameter are applied less to ocean-going ships than to inland vessels. The attainable propeller diameter can be increased to 90% of the draught and more. However, this increases resistance and suction resulting from the tunnel.
3. *Raising the counter* shortens the length of the waterline. This can increase the resistance. Relatively high counters are found on most banana carriers, which nearly always have limited draughts and relatively high power outputs.
4. *Extending the propeller below the keel line* is sometimes employed on destroyers and other warships, but rarely on cargo ships since the risk of damaging the propeller is too great.
5. *Increasing the draught* to accommodate a greater propeller diameter is often to be recommended, but not always possible. This decreases C_B and

the resistance. The draught can also be increased by a 'submarine keel'. Submarine keels, bar keels and box keels are found on trawlers, tugs and submarines.

6. *Kort nozzles* are only used reluctantly on ocean-going ships due to the danger of floating objects becoming jammed between the propeller and the inside of the nozzle. 'Safety nozzles' have been developed to prevent this. Kort nozzles also increase the risk of cavitation.
7. *Surface-piercing propellers* have been found in experiments to have good efficiency (Strunk, 1986; Miller and Szantyr, 1998), and are advocated for inland vessels, but no such installation is yet known to be operational.

Sterns for broad, shallow ships

High B/T ratios lead to large waterline run angles. The high resistance associated with a broad stern can be reduced by:

1. Small C_B and a small C_{WP}. Thus a greater proportion of the ship's length can be employed to taper the stern lines.
2. Where a local broadening of the stern is required, the resistance can be minimized by orientating the flowlines mainly along the buttock lines; i.e. the buttocks can be made shallow, thus limiting the extent of separated flow.
3. Where the stern is broad, a 'catamaran stern' (Fig. 2.33) with two propellers can be more effective, in terms of resistance and hull efficiency, than the normal stern form. At the outer surfaces of the catamaran stern the water is drawn into the propeller through small (if possible) waterline angles. The water between the propellers is led largely along the buttock lines. Hence it is important to have a flat buttock in the midship plane. Power requirements of catamaran sterns differ greatly according to design.

On broad ships, the normal rudder area is no longer sufficient in relation to the lateral plane area. This is particularly noticeable in the response to helm. It is advisable to relate the rudder area to the midship section area A_M. The rudder area should be at least 12% of A_M (instead of 1.6% of the lateral plane area). This method of relating to A_M can also be applied to fine ships.

In many cases it is advisable to arrange propeller shafts and bossings converging in the aft direction instead of a parallel arrangement.

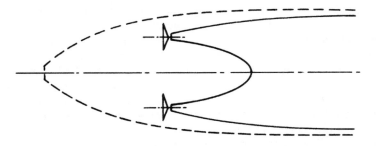

Figure 2.33 Catamaran stern. Waterplane at height of propeller shafts

2.8 Propeller clearances

The propeller blades revolving regularly past fixed parts of the ship produce hydrodynamic impulses which are transmitted into the ship's interior via both the external shell and the propeller shaft. The pressure impulses decrease the further the propeller blade tips are from the ship's hull and rudder. The 'propeller clearance' affects:

1. The power requirement.
2. Vibration-excitation of propeller and stern.
3. The propeller diameter and the optimum propeller speed.
4. The fluctuations in torque.

Vibrations may be disturbing to those on board and also cause fatigue fractures.

Clearance sizes

Propeller clearances have increased over time due to vibration problems (more power installed in lighter structures). High-skew propellers can somewhat counteract these problems since the impulses from the blade sections at different radii reach the counter at different times, reducing peaks. The pressure impulses increase roughly in inverse ratio to the clearance raised to the power of 1.5. The clearances are measured from the propeller contours as viewed from the side (Fig. 2.34). Where the propeller post is well rounded, the clearance should be taken from the idealized stern contour—the point of intersection of the outer shell tangents. The clearances in Fig. 2.34 are adequate unless special conditions prevail. A normal cargo ship without heel has a gap of 0.1–0.2 m between lower blade tip and base-line.

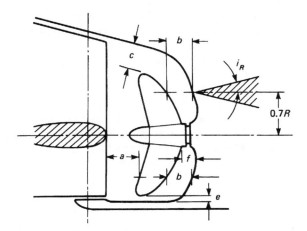

Figure 2.34 Propeller clearances; Det Norske Veritas recommendations for single-screw ships:

$a > 0.1D$	Horizontal to the rudder
$b > (0.35 - 0.02Z)D$	Horizontal to the propeller post
0.27D for four-bladed propellers	
$c > (0.24 - 0.01Z)D$	Vertical to the counter
0.20D for four-bladed propellers	
$e > 0.035D$	Vertical to the heel

Recommendations by Vossnack

The necessary propeller clearance for avoiding vibrations and cavitation is not a function of the propeller diameter, but depends primarily on the power and wake field and on a favourable propeller flow. Accordingly for single-screw ships the propeller clearance to the counter should be at least $c \approx 0.1$ mm/kW and the minimum horizontal distance at $0.7Rb \approx 0.23$ mm/kW.

Recommendations for twin-screw ships

$$c > (0.3 - 0.01Z) \cdot D \text{ according to Det Norske Veritas}$$

$$a > 2 \cdot (A_E/A_0) \cdot D/Z \text{ according to building regulations for German}$$
$$\text{naval vessels (BV 41)}$$

Here, Z is the number of propeller blades and A_E/A_0 the disc area ratio of the propeller.

These recommendations pay too little attention to important influences such as ship's form (angle of run of the waterlines), propulsion power and rpm. The clearances should therefore be examined particularly closely if construction, speed or power are unusual in any way. If C_B is high in relation to the speed, or the angle of run of the waterlines large or the sternpost thick, the clearance should be greater than recommended above.

The disadvantages of large clearances

1. Vertical clearances c and e:
   Relatively large vertical clearances limit the propeller diameter reducing the efficiency or increase the counter and thus the resistance.
2. Horizontal clearances a, b, f:
   A prescribed length between perpendiculars makes the waterlines more obtuse and increases the resistance. Against that, however, where the gap between propeller post and propeller is increased, the suction diminishes more than the accompanying wake, and this improves the hull efficiency $\eta_H = (1 - t)/(1 - w)$. This applies up to a gap of around two propeller diameters from the propeller post.
3. Distance from rudder a:
   Increasing the gap between rudder and propeller can increase or decrease power requirements. The rudder affects the power requirement in two ways, both of which are diminished when the gap increases. The result of this varies according to power and configuration. The effects are:
   (a) Fin effect, regaining of rotational energy in the slipstream.
   (b) Slipstream turbulence.

Summary: propeller clearances

Large clearances reduce vibrations. Small clearances reduce resistance: this results in a lower counter and a propeller post shifted aft. With regard to propulsion:

c and e should be small (to accommodate greater propeller diameter)

a and e should be small (possible regain of rotational energy at rudder section) *b* and *f* should be large (good hull efficiency η_H)

So the clearances *a*, *c* and *e* should be carefully balanced, since the require-ments for good vibration characteristics and low required output conflict. Only a relatively large gap between the propeller forward edge and the propeller post improves both vibration characteristics and power requirements—despite an increase in resistance.

Rudder heel

The construction without heel normally found today (i.e. open stern frame) has considerable advantages over the design with rudder heel:

1. Lower resistance (no heel and dead wood; possibility to position the counter lower).
2. Fewer surfaces to absorb vibration impulses.
3. Cheaper to build.

If a heel is incorporated after all, rounding off the upper part will decrease vibration (Fig. 2.35). For stern tunnels, the gap to the outer shell is normally smaller. Here, the distance between the blade tips and the outer shell should not change too quickly, i.e. the curvature of the outer shell should be hollow and the rounding-off radius of the outer shell should be greater than the propeller radius.

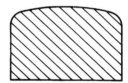

Figure 2.35 Rounded-off upper part of rudder heel

Taking account of the clearances in the lines design

To plot the clearances, the propeller silhouette and the rudder size must be known. Neither of these is given in the early design stages. Until more precise information is available, it is advisable to keep to the minimum values for the

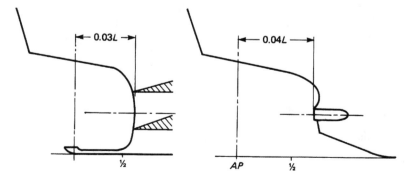

Figure 2.36 AP minimum distances between propeller post and aft perpendicular

distance between propeller post and aft perpendicular (Fig. 2.36). Vibrations can be reduced if the outer shell above the propeller is relatively stiff. This particular part of the outer shell can be made 1.8 times thicker than the surrounding area. Intermediate frames and supports add to the stiffening.

2.9 The conventional method of lines design

Lines design is to some extent an art. While the appearance of the lines is still important, today other considerations have priority. Conventionally, lines are either designed 'freely', i.e. from scratch, or distorted from existing lines, see Section 2.10.

The first stage in free design is to design the sectional area curve. There are two ways of doing this:

1. Showing the desired displacement as a trapezium (Fig. 2.37). The sectional area curve of the same area is derived from this simple figure.
2. Using an 'auxiliary diagram' to plot the sectional area curve.

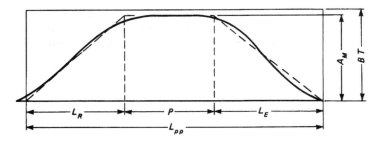

Figure 2.37 Design of sectional area curve (using a trapezium)

Sectional area curve using trapezium method

The ratio of the trapezium area to the rectangle with height A_M corresponds to the prismatic coefficient C_P. C_B is the ratio of this area to that of the rectangle with height $B \cdot T$. The length of the trapezium area is L_{pp}. The midship section area $A_M = B \cdot T \cdot C_M$ represents the height of the trapezium. The sectional area curve must show the desired displacement and centre of buoyancy.

The longitudinal centre of buoyancy can be determined by a moment calculation: it is also expressed in terms of the different coefficients of the fore and aftbodies of the ship. The geometric properties of the trapezium give:

Length of run $L_R = L_{pp}(1 - C_{PA})$

Length of entrance $L_E = L_{pp}(1 - C_{PF})$

C_{PA} is the prismatic coefficient of the aft part and C_{PF} is the prismatic coefficient of the fore part of the ship.

Recommendations for the length of run are:

(Baker) $L_R = 4.08\sqrt{A_M}$

(Alsen) $L_R = 3.2\sqrt{B \cdot T}/C_B$

Older recommendations for the entrance length are:

$$L_E = 0.1694 \cdot V^2 \quad V \text{ in kn}$$

$$= 0.64 \cdot V^2 \quad V \text{ in m/s}$$

$$= 6.3 \cdot F_n^2 \cdot L_{pp}$$

(Alsen) $L_E = 0.217 \cdot V^2$ with V in kn

Alsen's recommended values relate to the lengths of entrance and run up to the parallel middle body, i.e. they extend beyond the most sharply curved area of the sectional area curve. Recommendations such as these for entrance and run lengths can only be adhered to under certain conditions. If the three basic components of the trapezium—run, parallel middle body and entrance—and the main data—∇, L, B, T, and centre of buoyancy—are all fixed, there is little room for variation. In practice, it is only a matter of how the trapezium will be 'rounded' to give the same area. The lines designer may get the impression that a somewhat different sectional area curve would produce better faired lines. He should find a compromise, rather than try to make a success at all costs of the first sectional area curve. The wavelength (as a function of the water velocity) is extended at the bow by the increase in the water velocity caused by the displacement flow. On the other hand, the finite width of the ship makes the distance covered between stagnation point and shoulder longer than the corresponding distance on the sectional area curve. In modern practice, the shape of the forward shoulder is determined using CFD (see Section 2.11) to obtain the most favourable wave interaction.

Where there is no parallel middle body, the design trapezium becomes a triangle. This can be done for $F_n \approx 0.3$ (Fig. 2.38). The apex of the triangle must be higher than the midship section area on the diagram.

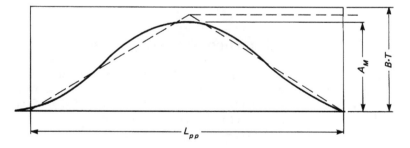

Figure 2.38 Sectional area curve for $F_n \approx 0.3$

Sectional area curves using design diagrams

Design diagrams of this kind are common in the literature, e.g. Lap (1954). Two alternatives are presented in this diagram: buoyancy distribution according to Lap and buoyancy distribution using the Series 60 model of the David Taylor Model Basin. The diagram shows the individual sectional areas from 0 to 20 as a function of the C_P as percentages of the midship sectional area. Different prismatic coefficients can be adopted for the forward and aftbodies.

The possibility of taking different prismatic coefficients for forward and aftbodies enables the longitudinal position of the centre of buoyancy to be varied independently. A precise knowledge of the buoyancy distribution is not absolutely essential to determine the centre of buoyancy. It usually suffices to know the fullness of fore and aftbodies to derive a centre of buoyancy sufficiently precise for lines design purposes. Equations for this are given in Section 2.10 on linear or affine distortion of ships' lines.

The criteria of the desired centre of buoyancy position and C_B are then used to form separate block coefficients for the fore and aftbodies, from which are derived the fore and aft prismatic coefficients to be entered in the diagram. Designing the sectional area curves using diagrams is preferable to the method using simple mathematical basic forms, since the sectional area curves taken from the design diagrams usually agree better with the lines, and thus accelerate the whole process. If it proves difficult to co-ordinate the lines with the sectional area curve, obtaining good lines should be given priority. When deviating from the sectional area curve, however, the displacement and its centre of buoyancy must always be checked. We presume that the conventional lines design procedure is known and will only highlight certain facts at this point.

The tolerances for displacement and centre of buoyancy are a function of ship type and the margins allowed for in the design. If the design is governed by a freeboard calculation, the displacement tolerance should be about ±0.5% at a 1–2% weight margin. A longitudinal centre of buoyancy tolerance of ±0.3%L_{pp} is acceptable. The associated difference in trim is approximately two-thirds of this. The vertical centre of buoyancy is not usually checked during the lines design.

Stability should be checked after the first fairing of the CWL (or a waterplane near the CWL). The transverse moment of inertia of the waterplane is roughly estimated. $\overline{BM}$ is obtained by dividing the value I_T by the nominal value for the displacement. To get $\overline{KM}$, a value for $\overline{KB}$ is added to $\overline{BM}$ using approximate formulae. The transverse moment of inertia of the waterplane is described in Section 1.2.

2.10 Lines design using distortion of existing forms

When designing the lines by distorting existing forms it usually suffices to design the underwater body and then add the topside in the conventional way. The bulbous bow is also often added conventionally. Thus a knowledge of the conventional methods is necessary even in distortion procedures.

Advantages of distortion over conventional procedures

1. Less work: there is no need to design a sectional area curve. Even where there is a sectional area curve, no checking of its concurrence with the lines is required.
2. It gives a general impression of many characteristics of the design before this is actually completed. Depending on the procedure applied, it may be possible, for example, to derive the value $\overline{KM}$.

Distortion methods

Existing forms with other dimensions and characteristics can be distorted in various ways:

1. Distorting lines given by drawings or tables of offset, by multiplying the offsets and shifting sectional planes such as waterlines, sections or buttocks.
   (a) Simple affine distortion, where length, width and height offsets are each multiplied by a standard ratio. If the three standard ratios are equal, geometric similarity is kept.
   (b) Modified affine distortion, where the simple affine distortion method is applied in a modified, partial or compound form.
   (c) Non-affine distortion, where the standard ratio can vary continuously in one or several directions.
2. Distorting lines given by mathematical equations.

Closed-form equations to represent the surfaces of normal ships are so complicated as to make them impracticable. The mathematical representation of individual surface areas using separate equations is more simple. For each boundary point belonging to two or more areas, i.e. which is defined by two or more equations, the equations must have the following points of identity: to avoid discontinuities, the ordinate values (half-widths) must be identical. To avoid a knuckle, the first derivatives with respect to x and z must be identical. The second derivatives should also be identical for good fairing. Whether this is required, however, depends on other conditions. CAD programs with 'graphical editors' help today to distort lines to the desired form. The following describes some of the distortion methods of the first group, distortion by multiplying offsets and shifting sectional planes, Schneekluth (1959).

(A) Linear or affine distortion (multiplication of offsets)

Affine distortion is where all the dimensions on each co-ordinate axis are changed proportionally. The scaling factors can be different for the three axes. As length L, width B, and draught T can be changed arbitrarily, so too can the ratios of these dimensions be made variable, e.g. L/B, B/T, ∇/L^3. Block coefficients, centres of buoyancy and waterplanes and the section character all remain unchanged in affine distortion. Before using other methods, the outlines must be affinely distorted to the desired main dimensions. In many cases, linear distortion is merely the preliminary stage in further distortion processes. It is not essential that fore and aftbodies be derived from the same ship.

Relations between the centre of buoyancy and the partial block coefficients of forward and aftbodies

With conventional lines, the relationships between the block coefficient C_B, the partial block coefficients C_{BF} (forebody), C_{BA} (aftbody) and the centre of buoyancy are more or less fixed. This is not a mathematical necessity, but can be expected in a conventional design. Suitable C_{BF} and C_{BA} are chosen to attain both the desired overall C_B and the desired centre of buoyancy. The following equation is used for this:

$$C_B = \frac{C_{BF} + C_{BA}}{2}$$

The following relationships were derived statistically.

Distance of centre of buoyancy before midship section

For cargo ships with $C_M > 0.94$:

$$\text{lcb}[\%L] = (C_{BF} - 0.973 \cdot C_B - 0.0211) \cdot 44$$

Rearranging these formulae gives:

$$C_{BF,BA} = C_B \pm \left(0.0211 + \frac{\text{lcb}}{44} - 0.027 \cdot C_B\right)$$

The midship section area coefficient C_M is neglected in this formula. Where C_M has arbitrary value:

$$\text{lcb}[\%L] = (C_{BF} - 0.973 \cdot C_B)\frac{43}{C_M} - 0.89$$

This produces after rearrangement:

$$C_{BF,BA} = C_B \pm (\text{lcb} + 0.89)\frac{C_M}{43} - 0.027 \cdot C_B$$

The error is $\Delta\text{lcb} < 0.1\%L$. The corresponding change in trim is $\Delta t < 0.07\%L$.

These equations apply to ships without bulbous bows. Ships with bulbs can be determined by estimating the volume of the bulb and then making allowance for it in a moment calculation.

Combining different designs

If the forward and aftbodies are derived from designs with different C_M, the results will be a sectional area curve in which the fore and aftbodies differ in height (Fig. 2.39). The lines in the midships area are usually combined by fairing by hand, a procedure involving little extra work. In any case, the midship section area normally has to be redesigned, since affine distortion using various factors for width and draught makes a quarter circle bilge into a quarter ellipse. Normally, however, a quadrant or hyperbolic bilge line is described.

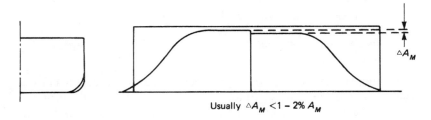

Usually $\Delta A_M < 1 - 2\% A_M$

Figure 2.39 Combining non-coherent sections

Requirements for further distortion procedures

All of the following methods are based on two conditions:

1. There is a choice between using a whole basis ship or two halves of different basis ships.

2. The first step is always linear distortion to attain the desired main dimensions. This is usually simply a case of converting the offsets and, if necessary, the waterline and section spacings.

Only a few of several existing distortion methods are mentioned here. These can easily be managed without the aid of computers, and have proved effective in practice. In the associated formulae, the basis ship (and the already linearly distorted basis ship) is denoted by the suffix v and the project to be designed by the suffix p.

(B) Interpolation (modified affine distortion)

The interpolation method offers the possibility of interpolation between the offsets of two forms, i.e. of seeking intermediate values in an arbitrary ratio (Fig. 2.40). The offsets to be interpolated must be of lines which have already been distorted linearly to fit the main dimensions by calculation. The interpolation can be graphical or numerical. In graphical interpolation, the two basis ships (affinely distorted to fit the main dimensions) are drawn in section and profile. The new design is drawn between the lines of, and at a constant distance from, the basis ships. One possible procedure in analytical interpolation is to give the basis ship an 'auxiliary waterplane subdivision' corresponding to that of the new design. For example if, using metric waterplane distances as a basis, the new design draught is 9 m, the draught of each of the two basis ships must be subdivided into nine equal distances. The half-widths are taken on these auxiliary waterplanes and multiplied in the ratio of new design width to basis ship width. This completes the first stage of affinely distorting both basis ships to fit the main dimensions of the new design. The offsets must now be interpolated. The procedure is the same for the side elevations. When interpolating, attention should be paid to the formation of the shoulders. A comparison of the new design sectional area curves with those of the basis ships shows that:

1. The fineness of the shoulders may be less marked in the new design than in the basis ships. A pronounced shoulder in the forebody can be of advantage if in the correct position.

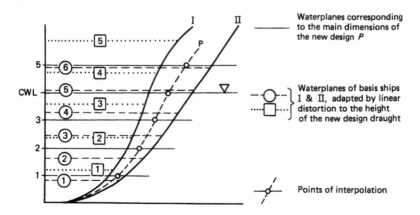

Figure 2.40 Lines design using interpolation method

2. Two shoulders can form if the positions of the shoulders in the two designs differ greatly (Fig. 2.41). Interpolations should therefore only be used with designs with similar shoulder positions.

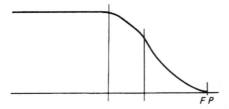

Figure 2.41 Possible formation of two shoulders through interpolation

The new displacement corresponds to the interpolation ratio:

$$\nabla_p = \nabla_{v1} + (\nabla_{v2} - \nabla_{v1}) \cdot x$$

Here, x represents the actual change in width, expressed as the ratio of the overall difference in width of the two design. The displacements ∇_{v1} and ∇_{v2} relate to the affinely distorted designs.

(C) Shift of design waterplane (modified affine distortion)

The normal distortion procedure considers the submerged part of the hull below the CWL. In the CWL shifting method, the draught of a basis ship and its halves is altered, i.e. a basis for the distortion is provided by that (either larger or smaller) part of the hull which is to be removed due to the new position of the CWL. Thus, C_B which decreases as the ship emerges progressively can be altered. The basis ship draught which gives the desired fullness can be read off directly from the normal position of C_B on the graph. Up to this draught, the sectional form of the design is used. Only its height is affinely distorted to the required new design draught; C_B remains unchanged. The new displacement is $\nabla_p = C_{Bv} \cdot L_p \cdot B_p \cdot T_p$ where C_{Bv} is the block coefficient changed by the CWL shift. Since C_B cannot be read very precisely from the graph, it is advisable to introduce the more precisely determined displacement of the design and determine C_B from that. Then

$$\nabla_p = \frac{\nabla_v}{L_v \cdot B_v \cdot T_v} \cdot L_p \cdot B_p \cdot T_p$$

Even without the hydrostatic curves, the change in fullness of the design can be estimated as a function of the draught variation.

A change in C_B changes other characteristics:

1. *Forebody*: A flared stem alters L_{pp}. The stem line should be corrected accordingly (Fig. 2.42).
2. *Aftbody*: There is a change in the ratio of propeller well height to draught. A change of this kind can be used to adapt the outline to the necessary propeller diameter or to alter transom submergence. L_{wl} changes, L_{pp} does not (Fig. 2.43).

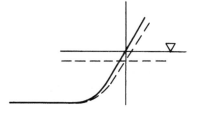

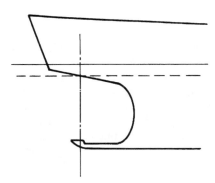

Figure 2.42 Correction to forward stem in the case of CWL shift

Figure 2.43 Effect of CWL shift at the stern

Such alterations to fore and aftbodies are usually only acceptable to a limited extent. Hence the CWL should only be shifted to achieve small changes in C_B.

(D) Variation of parallel middle body (modified affine distortion)

An extensively applied method to alter C_B consists of varying the length of the parallel middle body (Fig. 2.44). While the perpendiculars remain fixed, the section spacing is varied by altering the distances of the existing offset ordinates from the forward and aft perpendiculars in proportion to the factor K.

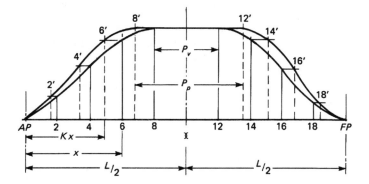

Figure 2.44 Variation in section spacings with change in length P of parallel middle body. Basis design P_v; new design P_p; change ΔP

The resulting new displacement can be determined exactly:

$$\nabla_p = \nabla_v \frac{(L_p - \Delta P) \cdot B_p \cdot T_p}{L_v \cdot B_v \cdot T_v} + \Delta P \cdot B_p \cdot T_p \cdot C_M$$

This formula also takes account of the simultaneous changes in L, B and T due to the linear distortion. If the ship has already been linearly distorted, the formula simplifies to:

$$\nabla_p = \nabla_v \frac{L - \Delta P}{L} + \Delta P \cdot B \cdot T \cdot C_M$$

In practice, this procedure is used primarily to increase the length of the parallel middle body and hence the fullness. Similarly, the fullness can be diminished by shortening the parallel middle body. If more length is cut away than is available in the parallel form, the ship will have a knuckle. This can affect sectional area curves, waterlines diagonals and buttocks. If moderate, this knuckle can be faired out. The shift in the positions of the shoulders and the individual section spacing can be determined using a simple formula. This assumes that the basis design has already been linearly distorted to the main dimensions of the new design. The length of the new parallel middle body is:

$$P_p = \Delta P + P_v \cdot K$$

The factor for the proportional change in all section spacings from the forward and aft perpendicular is:

$$K = \frac{L - \Delta P}{L} \qquad \Delta P = (1 - K)L$$

From geometrical relationships the resulting fullness is then:

$$C_{Bp} = \frac{C_{Bv}(L - \Delta P) \cdot B \cdot T + \Delta P \cdot B \cdot T \cdot C_M}{L \cdot B \cdot T}$$

By substituting ΔP and rearranging:

$$K = \frac{C_M - C_{Bp}}{C_M - C_{Bv}}$$

This gives factor K for the distance of the sections from the perpendiculars as a function of basis and proposed fullness and dependent on a common C_M.

This procedure can also be applied to each half of the ship separately, so that not only the size, but also the position of the parallel middle body can be changed. If fore and aftbodies are considered separately, the formula for one ship half is:

$$\Delta P = (1 - K)\frac{L}{2}$$

The block coefficients C_{Bp} and C_{Bv} of the corresponding half are to be inserted for K in the above formulae. The propeller aperture, and particularly the distance between propeller post and aft perpendicular, changes proportionally to the variation in section spacing. This must be corrected if necessary.

(E) Shift of section areas using parabolic curve (non-affine distortion)

Of the many characteristic curves for shift of section (Fig. 2.45) is very simple to develop. The changes of displacement are simple and can be determined with sufficient precision. The shifts in the sections can be plotted as a quadratic parabola over the length. If the parabola passes through the perpendicular and station 10 (at half the ship's length), the section shift will cause a subsequent change in the length of the propeller aperture. The dimensions of the propeller aperture are fixed, however, and should not be changed greatly. Unwanted changes can be avoided by locating the zero point of the parabola at the propeller post. Alternatively, desired changes in the size of the propeller aperture can be achieved by choosing the zero point accordingly, and most easily by trial and error. The height s of the parabola, its characteristic dimension, can be calculated from the intended difference in displacement and C_B. As Fig. 2.44 showed, the change in displacement can be represented in geometrical terms. Based on the already linearly distorted basic design, this amounts, for one-half of the ship, to:

$$\Delta \nabla = K \cdot B \cdot T \cdot C_M \cdot s = K \cdot A_M \cdot s$$

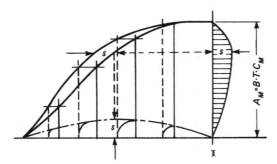

Figure 2.45 Distortion of sectional area curve using a parabolic curve

The change in C_B for one-half of the ship is:

$$\Delta C_B = \frac{2K \cdot s \cdot C_M}{L}$$

which gives the parabola height:

$$s = \frac{\Delta C_B \cdot L}{2K \cdot C_M} \quad \text{or} \quad s = \frac{\Delta \nabla}{K \cdot A_M}$$

$K \approx 0.7$ for prismatic coefficients $C_P < 0.6$. $K \approx 0.7 - (C_P - 0.6)^2 \cdot 4.4$ for $C_P \geq 0.6$. When using this procedure, it is advisable to check the change in displacement as a function of the parabola height s by distorting the sectional area curve. Only when the value s has been corrected if necessary, should the lines be carried forward to the new ship. Of all the methods described, this section shifting is the most universally applicable.

(F) Shifting the waterlines using parabolic curves (non-affine distortion)

As with the longitudinal shifting of sections using parabolic curves a similar procedure can be applied to shift the waterlines vertically (Fig. 2.46). There are two different types of application here:

1. Change in displacement and fullness with a simultaneous, more or less distinct change in the character of the section.
2. Change in section form and waterplane area coefficient with constant displacement.

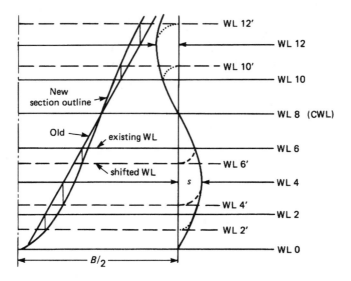

Figure 2.46 Effect of shift of waterlines on character of section

In both cases, the base-line and one waterline, e.g. the CWL, maintain their positions. The intermediate waterlines are displaced using the parabolic curve. As the diagram shows, waterlines above and below this line shift in the opposite direction. If the two parabola sections above and below the zero point are of different sizes, the curve must be faired at the turning-point. Even if this means that the waterplanes are shifted into the area above the CWL, the uppermost part of the flared section and the deck strake must still be designed in the conventional way. This method also involves varying the ratio of propeller aperture height to draught. However, the resulting changes in propeller aperture height are not as marked as those produced by linear shifting of the CWL. Shifting the waterlines naturally changes the vertical distances of the offsets for each section, and consequently the fullness and the character of the section as well. This method is therefore suitable for changing V sections to U sections, and vice versa. Choosing a suitable zero point of shift, the 'fixed' waterplane, allows distortion to be carried out with fullness unchanged with the sole purpose of changing the character of the section. Where the zero point of shift is situated in the CWL, i.e. the height of the displacement parabola is from the basis line to the CWL, the change in displacement is

$$\Delta \nabla = s \cdot K \cdot B \cdot L \cdot C_{WL}$$

and the associated change in fullness:

$$\Delta C_B = s \cdot K \cdot \frac{C_{WP}}{T}$$

where s is the greatest shift distance (largest ordinate in the shift parabola) and K is a factor dependent on the section character and the waterplane area curve and therefore on the ratio C_{WP}/C_B.

For absolutely vertical sections $C_{WP}/C_B = 1$ and $K = 0$. Normal cargo ships have a K value of 0.4–0.5. Using the curve of waterplane areas, K can be established for an existing ship by trial and error and then used in calculating the final distortion data. However, this is only possible if a complete ship is taken, since the hydrostatic curves only contain waterplane area specifications for the ship as a whole, not for the fore and aftbody separately.

Summary of areas of applications of the methods described

(A) Linear distortion and combination of different fore and aftbodies

Linear distortion is only possible if the basis ship has the desired C_B and centre of buoyancy, or if these are attainable by combining two suitable ships' halves.

(B) Interpolation

This method can result in flattening of shoulders. This is usually unimportant in the aftbody.

(C) Shifting the CWL with linear vertical distortion

This effectively changes the heights of both counter and propeller aperture in the aftbody. Otherwise, it is only suitable for small changes in C_B ($\Delta C_{BF} \le \pm0.012$, $\Delta C_{BA} \le \pm0.008$).

(D) Varying parallel middle body with linear distortion of the ship's ends

The change in the nature of the lines deserves careful attention.

(E) Shifting the section spacing using parabolic curves

This is the most practical method. The displacement and the centre of buoyancy are determined using an empirical coefficient dependent on the form of the section area curve. The size of this coefficient changes only marginally.

(F) Shifting the waterlines using parabolic curves

This enables the section characteristics to be changed, i.e. V sections to be developed from U forms and vice versa. The displacement and centre of buoyancy are determined using an empirical conversion coefficient which depends on the form of the waterplane area curve and the ratio C_{WP}/C_B.

There are several other distortion methods in addition to those listed above. In methods with total or partial affine distortion (A–D), the displacement is determined precisely without trial and error. Only for non-affine distortion (E and F) does the displacement depend on choosing the correct empirical coefficient. Here, the distances of shifting must be checked before the lines are drawn by evaluating the sectional area and the waterplane area curves.

Further hints for work with distortion methods

1. A critical waterline angle in the aftbody can be exceeded if linear distortion reduces the ratio L/B of the new design and the waterline angle of the basis ship is close to the critical.
2. The procedures described can also be used in combination.

Initial stability for ships with distorted lines

It is often possible in these distortion methods to derive values for the stability of the new design from those of the basis ship. In affine distortion factors used for length, width and draught are different. Here, the relationships between the stability of the new design and the basis ship are:

$$\overline{KM}_p = \overline{BM}_v \frac{(B_p/B_v)^2}{T_p/T_v} + \overline{KB}_v \frac{T_p}{T_v}$$

This equation can also be applied in the general design procedure if, when determining the main dimensions, the B/T ratio is varied for reasons of stability. Hence in affine distortion there is no need to determine the stability using approximate formulae for C_{WP}. The same applies in modified affine distortion. If interpolating the offsets of two outlines affinely distorted to the desired main dimensions produces a new outline design, the stability of the two distorted basis designs can first be determined using the above formula and then interpolated for the new design. If the differences are small, $\overline{KB}$ and $\overline{BM}$ may be linearly interpolated with sufficient precision. Should the CWL remain unchanged in the second distortion stage, i.e. if the new CWL corresponds to the affine distortion of the basis form, or if one of the basis waterlines becomes the new CWL in the proposed design (using method (C)), the lateral moment of inertia of the waterplane can be converted easily using the following equation, provided hydrostatic curves are available:

$$I_{Tp} = I_{Tv} \left(\frac{B_p}{B_v}\right)^3 \left(\frac{L_p}{L_v}\right)$$

If the second stage is variation of the parallel middle body (using method (D)), both the waterplane area coefficient and the waterplane transverse moment of inertia can be derived directly:

$$C_{WP,p} = \frac{C_{WP,v} \pm \Delta P/L}{1 \pm \Delta P/L}$$

ΔP is the change in length of parallel middle body.

$$I_{Tp} = I_{Tv} \left(\frac{B_p}{B_v}\right)^3 \cdot \frac{L_p - \Delta P}{L_v} + \frac{\Delta P \cdot B_p^3}{12}$$

Both formulae are correct—without empirical coefficients.

If the second stage is the parabolic variation of the section spacing (using method (E)), C_{WP} can only be derived approximately from the basis—but more precisely than by using approximate formulae:

$$C_{WP,p} \approx C_{WP,v} + \frac{2}{3}(C_{B,p} - C_{B,v})$$

Waterline angle with distortions

In affine distortion the tangent of the waterline angle i changes inversely with the L/B ratio. For a change in C_B due to a change in the parallel middle body and additional affine distortion we have:

$$\tan i_p = \tan i_v \frac{(1 - C_{P,v})L_v/B_v}{(1 - C_{P,p})L_p/B_p}$$

A change in C_B caused by parabolic distortion of the section spacings and additional affine distortion produces different changes in the waterline angles over the length of the ship. In the area of the perpendicular:

$$\tan i_p \approx \tan i_v (L/20)/(L/20 \pm 0.4s)$$

where s is the greatest shift of the parabola.

2.11 Computational fluid dynamics for hull design

CFD (computational fluid dynamics) is used increasingly to support model tests. For example, in Japan no ship is built that has not been previously analysed by CFD. CFD is often faster and cheaper than experiments and offers more insight into flow details, but its accuracy is still in many aspects insufficient, especially in predicting power requirements. This will remain so for some time. The 'numerical towing tank' in a strict sense is thus still far away. Instead, CFD is used for pre-selection of variants before testing and to study flow details to gain insight into how a ship hull can be improved. The most important methods in practice are panel methods for inviscid flows and 'Navier–Stokes' solvers for viscous flows. For hull lines design, in practice the applications are limited to the ship moving steadily ahead. This corresponds to a numerical simulation of the resistance or propulsion model test.

Grids used in the computations must capture the ship geometry appropriately, but also resolve changes in the flow with sufficient fineness. Usually one attempts to avoid extreme angles and side ratios in computational elements. Depending on ship and computational method, grid generation may take between hours and days, while the actual computer simulation runs automatically within minutes or hours and does not constitute a real cost factor. The most complicated task is grid generation on the ship hull itself; the remaining grid generation is usually automated. This explains why the analysis of further form variants for a ship is rather expensive, while variations of ship speed are cheap. Usually ship model basins can generate grids and perform CFD simulations better and faster than shipyards. The reason is that ship model basins profit from economies of scale, having more experience and specially developed auxiliary computer programs for grid generations, while individual shipyards use CFD only occasionally.

Computation of viscous flows

The Navier–Stokes equations (conservation of momentum) together with the continuity equation (conservation of mass) fully describe the flow about a ship. However, they cannot be solved analytically for real ship geometries. Additionally, a numerical solution cannot be expected in the foreseeable future. Therefore time-averaged Navier–Stokes equations (RANSE) are used to solve the problem. Since the actual Navier–Stokes equations are so far removed from being solved for ships, we often say 'Navier–Stokes' when meaning RANSE. These equations relate the turbulent fluctuations (Reynolds stresses) with the time-averaged velocity components. This relationship can only be supplied by semi-empirical theories in a turbulence model. All known turbulence models are plagued by large uncertainties. Furthermore, none of the turbulence models in use has ever been validated for applicability near the water surface. The choice of turbulence model influences, for example, separation of the flow in the computational model and thus indirectly the inflow to the propeller, a fundamental result of viscous computations. Despite certain progress, a comparative workshop, N.N. (1994), could not demonstrate any consistently convincing results for a tanker hull and a Series-60 hull ($C_B = 0.6$).

Navier–Stokes solvers discretize the fluid domain around the ship up to a certain distance in elements (cells). Typical cell numbers in the 1990s were between 100 000 und 500 000. Finite element methods, finite difference methods or finite volume methods are employed, with the latter being most popular.

Consideration of both viscosity and free-surface effects (wave-making, dynamic trim and sinkage) requires considerably more computational effort. Therefore many early viscous flow computations neglected the free surface and computed instead the double-body flow around the ship at model Reynolds number. The term 'double-body flow' indicates that a mirror image of the ship hull at the waterline in an infinite fluid domain gives for the lower half, automatically due to symmetry, the flow about the ship and a rigid water surface. This approximation is acceptable for slow ships and regions well below the waterline. For example, the influence of various bilge radii, the flow at waterjet or bow-thruster inlets, or even the propeller inflow for tankers may be properly analysed by this simplification. On the other hand, unacceptable errors have to be expected for propeller inflow in the upper region for fast ships, e.g. naval vessels or even some containerships. Unfortunately, numerical errors which are usually attributed to insufficient grid resolution and questionable turbulence models, make computation of the propeller inflow for full hull forms too inaccurate in practice. However, integral values of the inflow like the wake fraction are computed with good accuracy. Thus the methods are usually capable of identifying the best of several aftbody variants in design projects. Also some flows about appendages have been successfully analysed. The application of these viscous flow methods remains the domain of specialists, usually located in ship model basins, consulting the design engineer.

Methods that include both viscosity and free-surface effects are at the threshold of practical application. They will certainly become an important tool for lines design.

Computation of inviscid flows

If viscosity is neglected—and thus of course all turbulence effects—the Navier–Stokes equations simplify to the Euler equations, which have to be solved together with the continuity equation. They are rather irrelevant for ship flows. If the flow is assumed in addition to be free of rotation, we get to the Bernoulli and Laplace equations. If only the velocity is of interest, solution of the Laplace equation suffices. The Laplace equation is the fundamental equation for potential flows about ships. In a potential flow, the three velocity components are no longer independent from each other, but are coupled by the abstract quantity 'potential'. The derivative of the potential in any direction gives the velocity component in this direction. The problem is thus reduced to the determination of one quantity instead of three unknown velocity components. Of course, this simplifies the computation considerably. The Laplace equation is linear. This offers the additional advantage that elementary solutions (source, dipole, vortex) can be superimposed to arbitrarily complex solutions. Potential flow codes are still the most commonly used CFD tools for ship flows.

These potential flow codes are based on boundary element methods, also called panel methods. Panel methods discretize a surface, where a boundary condition can be numerically enforced, into a finite number (typically 1000 to 3000) discrete collocation points and a corresponding number of panels. At the collocation points, any linear boundary condition can be enforced exactly by adjusting the element strengths. For ship flows, hull and surrounding water surface are covered by elements. The boundary condition on the hull is zero normal velocity, i.e. water does not penetrate the ship hull. As viscosity is neglected, the tangential velocity may still be arbitrarily large. The boundary condition on the water surface is derived from the Bernoulli equation and thus initially contains squares of the unknown velocities. This nonlinear condition is fulfilled iteratively as a sequence of linear conditions. In each iterative step, the wave elevation and the dynamic trim and sinkage of the ship, i.e. the boundaries of the boundary element method, are adjusted until the nonlinear problem is solved with sufficient accuracy. All other boundary conditions are usually automatically fulfilled. Such 'fully nonlinear methods' were state of the art by 1990. They were developed and used at Flowtech (Sweden), HSVA (Germany), MARIN (Holland) and the DTRC (USA). Today, these programs are also directly employed by designers in shipyards, Krüger (1997). All commercial programs are similarly powerful, differences in the quality of the results stem rather from the experience and competence of the user.

Panel methods have fundamental restrictions which have to be understood by the user. Disregarding viscosity introduces considerable errors in the aftbody. Thus the inflow to the propeller is not even remotely correct. Therefore, inviscid ship flow computations do not include propeller or rudder. The hull must be smooth and streamlines may not cross knuckles, as an ideal potential flow attains infinite velocity flowing around a sharp corner while a real flow will separate here. The solution in these cases is a generous rounding of the ship geometry. This avoids formal problems in the computations, but of course at the price of a locally completely different flow. Planing is also difficult to capture properly. Furthermore, none of the methods used in practice is capable of modelling breaking waves. This is problematic in the immediate vicinity of the bow for all ships, but also further away from the hull for catamarans if interference of the wave systems generated by the two hulls leads to local splashes. In this case, only linear and thus less accurate solutions can be obtained. In addition to these limitations from the underlying physical assumptions, there are practical limitations due to the available computer capacity. Slow ships introduce numerical difficulties if the waves—getting quadratically shorter with decreasing Froude number—have to be resolved by the grid.

The application of panel methods is thus typically limited to displacement ships with Froude numbers $0.15 < F_n < 0.4$. This interval fortunately covers almost all cargo and naval vessels. There are many publications presenting applications for various conventional ship forms (tanker, containerships, ferries), but also sailing yachts and catamarans with and without an air cushion, e.g. Bertram (1994), Bertram and Jensen (1994), Larsson (1994). By far the most common application of panel methods is the evaluation of various bow variants for pre-selection of the ship hull before model tests are conducted. The methods are not suited to predicting resistance, simply because wave breaking and viscous pressure resistance cannot be captured. Instead, one compares pressure distributions and wave patterns for various hull forms with comparable grids. This procedure has now become virtually a standard for designing bulbous bows.

Dynamical trim and sinkage are computed accurately by these methods and can serve, together with the computed wave pattern, as input for more sophisticated viscous flow computations.

Representation of results

CFD methods produce a host of data, e.g. velocities at thousands of points. This amount of data requires aggregation to a few numbers and display in suitable automatic plots both for quality control and evaluation of the ship hull.

The following displays are customary:

- Pressure distributions on the hull
  Colour plots of interpolated contour lines of pressures allow identification of critical regions. Generally, one strives for an even pressure distribution. Strong pressure fluctuations in the waterline correspond to pronounced wave troughs and crests, i.e. high wave resistance. Interpolation of pressures over the individual elements leads to a more realistic pressure pattern, however, the grid fineness determines the accuracy of this interpolation. Therefore, a plot of the grid should always accompany the pressure plot.
- Wave profile of the hull
  Ship designers are accustomed to evaluating a ship form from the wave profile on the hull, based on their experience with model tests. In a comparison of variants, the wave profiles show which form has the better wave systems interference, and thus the lower wave resistance. For clarity, CFD plots usually amplify the vertical co-ordinate by, for example, a factor of 5. Interpolation again gives the illusion of higher data density.
- Velocity plots on the hull
  Velocity plots give the local flow direction similar to tuft tests in model experiments. This is used for evaluating bulbous bows, but also for arranging bilge keels.

- Wave pattern

  Plots of contourlines of the wave elevation are mainly used for quality control. Reflections on the border of the computational domain and waves at the upstream border of the grid indicate that the grid was too small and the computation should be repeated with a larger grid. Typically but with no indication of numerical error, waves at the stern are higher than at the bow. This is due to larger run angles than entrance angles and the neglect of viscosity, which in reality reduces the waves at the aftbody.

- Perspective view of water surface

  Perspective views of the water surface, often with 'hidden-lines' or shading are popular, but have no value for designing better hull forms.

Often pressure, velocity and wave elevation are combined in one plot.

CFD reports should contain, as a minimum, the following information (Bertram, 1992):

Information for form improvement

1. Pressure contour lines (preferably in colour) in all perspectives needed to show the relevant regions. Oblique views from top and bottom have been proven as suitable.
2. Wave profile at hull with information on how the profile was interpolated and the vertical scale factor.
3. Velocity contribution at forebody showing the flow directions. The ship speed should be given as a reference vector.
4. An estimate of the relative change in resistance for comparison of variants versus a basis form.

Information for quality control

1. Plots of grids, especially on the hull, to provide a reference for the accuracy of interpolated results.
2. Information on the convergence of iterative solutions.
3. Plots of wave pattern to detect implausible results at the outer boundary of the computational domain or at the ship ends.

Generally, plots of the hull should contain main reference lines (CWL, sections) to facilitate the reference to the lines plan.

2.12 References

BERTRAM, V. (1992). CFD im Schiffbau. *Handbuch der Werften* Vol. **XXI**, Hansa, p. 19

BERTRAM, V. (1994). Numerische Schiffshydrodynamik in der Praxis. IFS-Report 545, Univ. Hamburg

BERTRAM, V. and JENSEN, G. (1994). Recent applications of computational fluid dynamics. *Schiffstechnik*, p. 131

DANCKWARDT, E. (1969). Ermittlung des Widerstandes von Frachtschiffen und Hecktrawlern beim Entwurf. *Schiffbauforschung*, p. 124

ECKERT, E. and SHARMA, S. (1970). Bugwülste für langsame, völlige Schiffe. *Jahrbuch Schiffbautechn. Gesellschaft*, p. 129

HÄHNEL, G. and LABES, K.-H. (1968). Systematische Widerstandsuntersuchungen für schnelle Frachtschiffe mit und ohne Bugwulst. *Schiffbauforschung*, p. 85

HOEKSTRA, M. (1975). Prediction of full scale wake characteristics. *International Shipbuilding Progress*, p. 204

HOLDEN, K. O., FAGERJORD, O. and FROSTAD, R. (1980). Early design-stage approach to reducing hull surface forces due to propeller cavitation. *Trans. SNAME* **88**, p. 403

HOYLE, J. W., CHENG, B. H., HAYS, B., JOHNSON, B. and NEHRLING, B. (1986). A bulbous bow design methodology for high-speed ships. *Trans. SNAME* **94**, p. 31

JENSEN, G. (1994). Moderne Schiffslinien. *Handbuch der Werften* Vol. **XXII**, p. 93

KERLEN, H. (1971). Entwurf von Bugwülsten für völlige Schiffe aus der Sicht der Praxis. *Hansa*, p. 1031

KRACHT, A. (1973). Theoretische und Experimentelle Untersuchungen für die Anwendung von Bugwülsten. Report 36, Forschungszentrum des Deutschen Schiffbaus, Hamburg

KRÜGER, S. (1997). Moderne hydrodynamische Entwurfsmethoden in der Werftpraxis. *Jahrbuch Schiffbautechn. Gesellschaft*

LAP, A. J. W. (1954). Diagrams for determining the resistance of single-screw ships. *International Shipbuilding Progress*, p. 179

LARSSON, L. (1994). CFD as a tool in ship design. In *N. N. (1994)*

MILLER, W. and SZANTYR, J. (1998), Model experiments with surface piercing propellers, *Schiffstechnik* **45**

N. N. (1994). *CFD Workshop Tokyo 1994*. Ship Research Institute

POPHANKEN, E. (1939). *Schiffbaukalender—Hilfsbuch der Schiffbauindustrie*. Deutsche Verlagswerke, Berlin

SCHNEEKLUTH, H. (1959). Einige Verfahren und Näherungsformeln zum Gebrauch beim Linienentwurf. *Schiffstechnik*, p. 130

STRUNK, H. (1986). Systematische Freifahrtsversuche mit teilgetauchten Propellern unter Druckähnlichkeit. Report 180, Forschungszentrum des Deutschen Schiffbaus, Hamburg

TZABIRAS, G. (1997). A numerical study of additive bulb effects on the resistance and self-propulsion of a full ship form. *Schiffstechnik*, p. 98

WURR, D. (1979). Heckwulst in vereinfachter Bauweise für Einschraubenschiffe. *Hansa*, p. 1796

3
Optimization in design

Most design problems may be formulated as follows: determine a set of design variables (e.g. number of ships, individual ship size and speed in fleet optimization; main dimensions and interior subdivision of ship; scantlings of a construction; characteristic values of pipes and pumps in a pipe net) subject to certain relations between and restrictions of these variables (e.g. by physical, technical, legal, economical laws). If more than one combination of design variables satisfies all these conditions, we would like to determine that combination of design variables which optimizes some measure of merit (e.g. weight, cost, or yield).

3.1 Introduction to methodology of optimization

Optimization means finding the best solution from a limited or unlimited number of choices. Even if the number of choices is finite, it is often so large that it is impossible to evaluate each possible solution and then determine the best choice. There are, in principle, two methods of approaching optimization problems:

1. Direct search approach
   Solutions are generated by varying parameters either systematically in certain steps or randomly. The best of these solutions is then taken as the estimated optimum. Systematic variation soon becomes prohibitively time consuming as the number of varied variables increases. Random searches are then employed, but these are still inefficient for problems with many design variables.
2. Steepness approach
   The solutions are generated using some information on the local steepness (in various directions) of the function to be optimized. When the steepness in all directions is (nearly) zero, the estimate for the optimum is found. This approach is more efficient in many cases. However, if several local optima exist, the method will 'get stuck' at the nearest local optimum instead of finding the global optimum, i.e. the best of *all* possible solutions. Discontinuities (steps) are problematic; even functions that vary steeply in one direction, but very little in another direction make this approach slow and often unreliable.

Most optimization methods in ship design are based on steepness approaches because they are so efficient for smooth functions. As an example consider the cost function varied over length L and block coefficient C_B (Fig. 3.1). A steepness approach method will find quickly the lowest point on the cost function, if the function $K = f(C_B, L)$ has only one minimum. This is often the case.

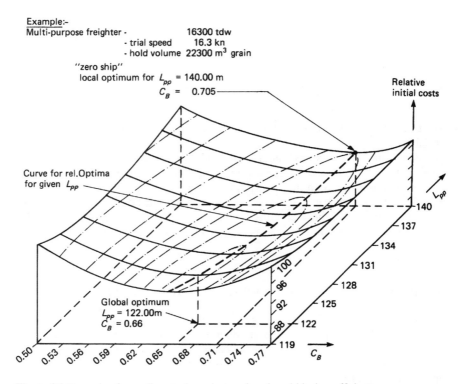

Example:-
Multi-purpose freighter -   16300 tdw
                    - trial speed   16.3 kn
                    - hold volume 22300 m³ grain

"zero ship"
local optimum for L_{pp} = 140.00 m
                    C_B =   0.705

Relative initial costs

Curve for rel.Optima for given L_{pp}

Global optimum
L_{pp} = 122.00m
C_B = 0.66

L_{pp}
140
137
134
131
128
125
122
119

100
96
92
88

C_B
0.50 0.53 0.56 0.59 0.62 0.65 0.68 0.71 0.74 0.77

Figure 3.1 Example of overall costs dependent on length and block coefficient

Repeating the optimization with various starting points may circumvent the problem of 'getting stuck' at local optima. One option is to combine both approaches with a quick direct search using a few points to determine the starting point of the steepness approach. Also repeatedly alternating both methods—with the direct approach using a smaller grid scale and range of variation each time—has been proposed.

A pragmatic approach to treating discontinuities (steps) assumes first a continuous function, then repeats the optimization with lower and upper next values as fixed constraints and taking the better of the two optima thus obtained. Although, in theory, cases can be constructed where such a procedure will not give the overall optimum, in practice this procedure apparently works well.

The target of optimization is the objective function or criterion of the optimization. It is subject to boundary conditions or constraints. Constraints may be formulated as equations or inequalities. All technical and economical relationships to be considered in the optimization model must be known and expressed as functions. Some relationships will be exact, e.g. $\nabla = C_B \cdot L \cdot B \cdot T$; others will only be approximate, such as all empirical formulae, e.g. regarding resistance or weight estimates. Procedures must be sufficiently precise, yet may not consume too much time or require highly detailed inputs. Ideally all variants should be evaluated with the same procedures. If a change of procedure is necessary, for example, because the area of validity is exceeded, the results of the two procedures must be correlated or blended if the approximated quantity is continuous in reality.

A problem often encountered in optimization is having to use unknown or uncertain values, e.g. future prices. Here plausible assumptions must be made. Where these assumptions are highly uncertain, it is common to optimize for several assumptions ('sensitivity study'). If a variation in certain input values only slightly affects the result, these may be assumed rather arbitrarily.

The main difficulty in most optimization problems does not lie in the mathematics or methods involved, i.e. whether a certain algorithm is more efficient or robust than others. The main difficulty lies in formulating the objective and all the constraints. If the human is not clear about his objective, the computer cannot perform the optimization. The designer has to decide first what he really wants. This is not easy for complex problems. Often the designer will list many objectives which a design shall achieve. This is then referred to in the literature as 'multi-criteria optimization', e.g. Sen (1992), Ray and Sha (1994). The expression is nonsense if taken literally. Optimization is only possible for one criterion, e.g. it is nonsense to ask for the best and cheapest solution. The best solution will not come cheaply, the cheapest solution will not be so good. There are two principle ways to handle 'multi-criteria' problems, both leading to one-criterion optimization:

1. One criterion is selected and the other criteria are formulated as constraints.
2. A weighted sum of all criteria forms the optimization objective. This abstract criterion can be interpreted as an 'optimum compromise'. However, the rather arbitrary choice of weight factors makes the optimization model obscure and we prefer the first option.

Throughout optimization, design requirements (constraints), e.g. cargo weight, deadweight, speed and hold capacity, must be satisfied. The starting point is called the 'basis design' or 'zero variant'. The optimization process generates alternatives or variants differing, for example, in main dimensions, form parameters, displacement, main propulsion power, tonnage, fuel consumption and initial costs. The constraints influence, usually, the result of the optimization. Figure 3.2 demonstrates, as an example, the effects of different optimization constraints on the sectional area curve.

Optimized main dimensions often differ from the values found in built ships. There are several reasons for these discrepancies:

1. *Some built ships are suboptimal*
   The usual design process relies on statistics and comparisons with existing ships, rather than analytical approaches and formal optimization. Designs found this way satisfy the owner's requirements, but better solutions, both for the shipyard and the owner, may exist. Technological advances, changes in legislation and in economical factors (e.g. the price of fuel) are reflected immediately in an appropriate optimization model, but not when relying on partially outdated experience. Modern design approaches increasingly

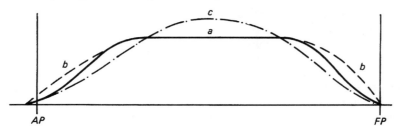

Figure 3.2 Changes produced in sectional area curve by various optimization constraints:
a is the basis form;
b is a fuller form with more displacement; optimization of carrying capacity with maximum main dimensions and variable displacement;
c is a finer form with the displacement of the basis form *a*, with variable main dimensions

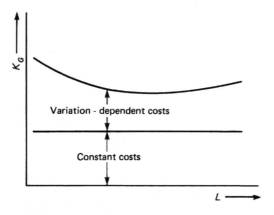

Figure 3.3 Division of costs into length-dependent and length-independent

incorporate analyses in the design and compare more variants generated with the help of the computer. This should decrease the differences between optimization and built ships.
2. *The optimization model is insufficient*
   The optimization model may have neglected factors that are important in practice, but difficult to quantify in an optimization procedure, e.g. seakeeping behaviour, manoeuvrability, vibrational characteristics, easy cargo-handling. Even for directly incorporated quantities, often important relationships are overlooked, leading to wrong optima, e.g.:
   (a) A faster ship usually attracts more cargo, or can charge higher freight rates, but often income is assumed as speed independent.
   (b) A larger ship will generally have lower quay-to-quay transport costs per cargo unit, but time for cargo-handling in port may increase. Often, the time in port is assumed to be size independent.
   (c) In reefers the design of the refrigerated hold with regard to insulation and temperature requirements affects the optimum main dimensions. The additional investment and annual costs have to be included in the model to obtain realistic results.

(d) The performance of a ship will often deteriorate over time. Operating costs will correspondingly increase, Malone *et al.* (1980), Townsin *et al.* (1981), but are usually assumed time independent.

The economic model may use an inappropriate objective function. Often there is confusion over the treatment of depreciation. This is not an item of expenditure, i.e. cash flow, but a book-keeping and tax calculation device, see Section 3.3. The optimization model may also be based on too simplified technical relationships. Most of the practical difficulties boil down to obtaining realistic data to include in the analysis, rather than the mechanics of making the analysis. For example, the procedures for weight estimation, power prediction and building costs are quite inaccurate, which becomes obvious when the results of different published formulae are compared. The optimization process may now just maximize the error in the formulae rather than minimize the objective.

The result of the optimization model should be compared against built ships. Consistent differences may help to identify important factors so far neglected in the model. A sensitivity analysis concerning the underlying estimation formulae will give a bandwidth of 'optimal' solutions and any design within this bandwidth must be considered as equivalent. If the bandwidth is too large, the optimization is insignificant.

A critical view on the results of optimization is recommended. But properly used optimization may guide us to better designs than merely reciprocating traditional designs. The ship main dimensions should be appropriately selected by a naval architect who understands the relationships of various variables and the pitfalls of optimization. An automatic optimization does not absolve the designer of his responsibility. It only supports him in his decisions.

3.2 Scope of application in ship design

Formal optimization of the lines including the bulbous bow even for fixed main dimensions is beyond our current computational capabilities. Although such formal optimization has been attempted using CFD methods, the results were not convincing despite high computational effort (Janson, 1997). Instead, we will focus here on ship design optimization problems involving only a few (less than 10) independent variables and rather simple functions. A typical application would be the optimization of the main dimensions. However, optimization may be applied to a wide variety of ship design problems ranging from fleet optimization to details of structural design.

In fleet optimization, the objective is often to find the optimum number of ships, ship speed and capacity without going into further details of main dimensions, etc. A ship's economic efficiency is usually improved by increasing its size, as specific cost (cost per unit load, e.g. per TEU or per ton of cargo) for initial cost, fuel, crew, etc., decrease. However, dimensional limitations restrict size. The draught (and thus indirectly the depth) is limited by channels and harbours. However, for draught restrictions one should keep in mind that a ship is not always fully loaded and harbours may be dredged to greater draughts during the ship's life. The width of tankers is limited by building and repair docks. The width of containerships is limited by the span of container bridges. Locks restrict all the dimensions of inland vessels. In addition, there are less obvious aspects limiting the optimum ship size:

1. The limited availability of cargo coupled to certain expectations concerning frequency of departure limits the size on certain routes.
2. Port time increases with size, reducing the number of voyages per year and thus the income.
3. The shipping company loses flexibility. Several small ships can service more frequently various routes/harbours and will thus usually attract more cargo. It is also easier to respond to seasonal fluctuations.
4. Port duties increase with tonnage. A large ship calling on many harbours may have to pay more port dues than several smaller ships servicing the same harbours in various routes, thus calling each in fewer harbours.
5. In container line shipping, the shipping companies offer door-to-door transport. The costs for feeder and hinterland traffic increase if large ships only service a few 'hub' harbours and distribute the cargo from there to the individual customer. Costs for cargo-handling and land transport then often exceed savings in shipping costs.

These considerations largely concern shipping companies in optimizing the ship size. Factors favouring larger ship size are (Buxton, 1976):

- Increased annual flow of cargo.
- Faster cargo-handling.
- Cargo available one way only.
- Long-term availability of cargo.
- Longer voyage distance.
- Reduced cargo-handling and stock-piling costs.
- Anticipated port improvements.
- Reduced unit costs of building ships.
- Reduced frequency of service.

We refer to Benford (1965) for more details on selecting ship size.

After the optimum size, speed, and number of ships has been determined along with some other specifications, the design engineer at the shipyard is usually tasked to perform an optimization of the main dimensions as a start of the design. Further stages of the design will involve local hull shape, e.g. design of the bulbous bow lines, structural design, etc. Optimization of structural details often involves only a few variables and rather exact functions. Söding (1977) presents as an example the weight optimization of a corrugated bulkhead. Other examples are found in Liu et al. (1981) and Winkle and Baird (1985).

For the remainder of the chapter we will discuss only the optimization of main dimensions for a single ship. Pioneering work in introducing optimization to ship conceptual design in Germany has been performed by the Technical University of Aachen (Schneekluth, 1957, 1967; Malzahn et al., 1978). Such an optimization varies technical aspects and evaluates the result from an economic viewpoint. Fundamental equations (e.g. $\nabla = C_B \cdot L \cdot B \cdot T$), technical specifications/constraints, and equations describing the economical criteria form a more or less complicated system of coupled equations, which usually involve nonlinearities. Gudenschwager (1988) gives an extensive optimization model for ro-ro ships with 57 unknowns, 44 equations, and 34 constraints.

To establish such complicated design models, it is recommended to start with a few relations and design variables, and then to improve the model step by step, always comparing the results with the designer's experience and understanding the changes relative to the previous, simpler model. This is necessary in a complicated design model to avoid errors or inaccuracies which cannot be clarified or which may even remain unnoticed without applying this stepwise procedure. Design variables which involve step functions (number of propeller blades, power of installed engines, number of containers over the width of a ship, etc.) may then be determined at an early stage and can be kept constant in a more sophisticated model, thus reducing the complexity and computational effort. Weakly variation-dependent variables or variables of secondary importance (e.g. displacement, underdeck volume, stability) should only be introduced at a late stage of the development procedure. The most economic solution often lies at the border of the search space defined by constraints, e.g. the maximum permissible draught or Panamax width for large ships. If this is realized in the early cycles, the relevant variables should be set constant in the optimization model in further cycles. Keane *et al.* (1991) discuss solution strategies of optimization problems in more detail.

Simplifications can be retained if the associated error is sufficiently small. They can also be given subsequent consideration.

3.3 Economic basics for optimization

Discounting

For purposes of optimization, all payments are discounted, i.e. converted by taking account of the interest, to the time when the vessel is commissioned. The rate of interest used in discounting is usually the market rate for long-term loans. Discounting decreases the value of future payments and increases the value of past payments. Individual payments thus discounted are, for example, instalments for the new building costs and the re-sale price or scrap value of the ship. The present value (discounted value) K_{pv} of an individual payment K paid l years later—e.g. scrap or re-sale value—is:

$$K_{pv} = K \cdot \frac{1}{(1+i)^l} = K \cdot \text{PWF}$$

where i is the interest rate. PWF is the present worth factor. For an interest rate of 8%, the PWF is 0.2145 for an investment life of 20 years, and 0.9259 for 1 year. If the scrap value of a ship after 20 years is 5% of the initial cost, the discounted value is about 1%. Thus the error in neglecting it for simplification is relatively small.

A series of constant payments k is similarly discounted to present value K_{pv} by:

$$K_{pv} = k \cdot \frac{(1+i)^l \cdot i}{(1+i)^l - 1} = k \cdot \text{CRF}$$

CRF is the capital recovery factor. The shorter the investment life, the greater is the CRF at the same rate of interest. For an interest rate of 8%, the CRF is 0.1018 for 20 years and 1.08 for 1 year of investment life.

The above formulae assume payment of interest at the end of each year. This is the norm in economic calculations. However, other payment cycles can easily be converted to this norm. For example, for quarterly payments divide i by 4 and multiply l by 4 in the above formulae.

For costs incurred at greater intervals than years, or on a highly irregular basis, e.g. large-scale repair work, an annual average is used. Where changes in costs are anticipated, future costs should be entered at the average annual level as expected. Evaluation of individual costs is based on present values which may be corrected if recognizable longer-term trends exist. Problems are:

1. The useful life of the ship can only be estimated.
2. During the useful life, costs can change with the result that cost components may change in absolute terms and in relation to each other. After the oil crisis of 1973, for example, fuel costs rose dramatically.

All expenditure and income in a ship's life can thus be discounted to a total 'net present value' (NPV). Only the cash flow (expenditure and income) should be considered, not costs which are used only for accounting purposes.

Yield is the interest rate i that gives zero NPV for a given cash flow. Yield is also called Discounted Cash Flow Rate of Return, or Internal Rate of Return. It allows comparisons between widely different alternatives differing also in capital invested. In principle, yield should be used as the economic criterion to evaluate various ship alternatives, just as it is used predominantly in business administration as the benchmark for investments of all kinds. The operating life should be identical for various investments then. Unfortunately, yield depends on uncertain quantities like future freight rates, future operating costs, and operating life of a ship. It also requires the highest computational effort as building costs, operating costs and income must all be estimated.

Other economic criteria which consider the time value of money include NPV, NPV/investment, or Required Freight Rate (= the freight rate that gives zero NPV); they are discussed in more detail by Buxton (1976). The literature is full of long and rather academic discussions on what is the best criterion. But the choice of the economic criterion is actually of secondary importance in view of the possible errors in the optimization model (such as overlooking important factors or using inaccurate relationships).

Discounting decreases the influence of future payments. The initial costs are not discounted, represent the single most important payment and are the least afflicted by uncertainty. (Strictly speaking, the individual instalments of the initial costs should be discounted, but these are due over the short building period of the ship.) The criterion 'initial costs' simplifies the optimization model, as several variation-independent quantities can be omitted. Initial costs have often been recommended as the best criterion for shipyard as this maximizes the shipyard's profit. This is only true if the price for various alternatives is constant. However, in modern business practice the shipyard has to convince the shipowner of its design. Then price will be coupled to expected cash flow.

In summary, the criterion for optimization should usually be yield. For a simpler approach, which may often suffice or serve in developing the optimization model, initial costs may be minimized.

Initial costs (building costs)

Building costs can be roughly classified into:

- Direct labour costs.
- Direct material costs (including services bought).
- Overhead costs.

Overhead costs are related to individual ships by some appropriate key, for example equally among all ships built at the accounting period, proportional to direct costs, etc.

For optimization, the production costs are divided into (Fig. 3.3):

1. *Variation-dependent costs*
   Costs which depend on the ship's form:
   (a) Cost of hull.
   (b) Cost of propulsion unit (main engine).
   (c) Other variation-dependent costs, e.g. hatchways, pipes, etc.
2. *Variation-independent costs*
   Costs which are the same for every variant, e.g. navigation equipment, living quarters, etc.

Buxton (1976) gives some simple empirical estimates for these costs.

Building costs are covered by own capital and loans. The source of the capital may be disregarded. Then also interest on loans need not be considered in the cash flow. The yield on the capital should then be larger than alternative forms of investment, especially the interest rate of long-term loans. This approach is too simple for an investment decision, but suffices for optimizing the main dimensions.

Typically 15–45% of the initial costs are attributable to the shipyard, the rest to outside suppliers. The tendency is towards increased outsourcing. Of the wages paid by the shipyard, typically 20% are allotted to design and 80% to production for one-of-a-kind cargo ships, while warships feature typically a 50:50 proportion.

Determining the variation-dependent costs

Superstructure and deckhouses are usually assumed to be variation-independent when considering variations of main dimensions. The variation-dependent costs are:

1. The hull steel costs.
2. The variation-dependent propulsion unit costs.
3. Those components of equipment and outfit which change with main dimensions.

The steel costs

The yards usually determine the costs of the processed steel in two separate groups:

1. The cost of the unprocessed rolled steel. The costs of plates and rolled sections are determined separately using prices per ton. The overall weight

is determined by the steel weight calculation. The cost of wastage must be added to this.

2. Other costs. These comprise mainly wages. This cost group depends on the number of man-hours spent working on the ship within the yard. The numbers differ widely, depending on the production methods and complexity of construction. As a rough estimate, 25–35 man-hours/t for containerships are cited in older literature. There are around 30–40% more man-hours/t needed for constructing the superstructure and deckhouses than for the hull, and likewise for building the ship's ends as compared with the parallel middlebody. The amount of work related to steel weight is greater on smaller ships. For example, a ship with $70\,000\,\text{m}^3$ underdeck volume needs 15% less manufacturing time per ton than a ship with $20\,000\,\text{m}^3$ (Kerlen, 1985).

For optimization, it is more practical to form 'unit costs per ton of steel installed', and then multiply these unit costs by the steel weight. These unit costs can be estimated as the calculated production costs of the steel hull divided by the net steel weight. Kerlen (1985) gives the specific hull steel costs as:

$$k_{St}[\text{MU/t}] = k_0 \cdot \left(\frac{4}{\sqrt[3]{L/\text{m}}} + \frac{3}{L/\text{m}} + 0.2082 \right)$$

$$\cdot \left(\frac{3}{2.58 + C_B^2} - 0.07 \cdot \frac{0.65 - C_B}{0.65} \right)$$

k_0 represents the production costs of a ship 140 m in length with $C_B = 0.65$. The formula is applicable for ships with $0.5 \le C_B \le 0.8$ and $80\,\text{m} \le L \le 200\,\text{m}$. The formula may be modified, depending on the material costs and changes in work content.

Propulsion unit costs

For optimization of main dimensions, the costs of the propulsion plant may be assumed to vary continuously with propulsion power. They can then be obtained by multiplying propulsion power by unit costs per unit of power. A further possibility is to use the catalogue prices for engines, gears and other large plant components in the calculation and to take account of other parts of the machinery by multiplying by an empirical factor. Only those parts which are functions of the propulsion power should be considered. The electrical plant, counted as part of the engine plant in design—including the generators, ballast water pipes, valves and pumps—is largely variation-independent.

The costs of the weight group 'equipment and outfit'

Whether certain parts are so variation-dependent as to justify their being considered depends on the ship type. For optimization of initial costs, the equipment can be divided into three groups:

1. Totally variation-independent equipment, e.g. electronic units on board.
2. Marginally variation-dependent equipment, e.g. anchors, chains and hawsers which can change if in the variation the classification numeral changes. If

variation-dependence is not pronounced, the equipment in question can be omitted.
3. Strongly variation-dependent equipment, e.g. the cost of hatchways rises roughly in proportion to the hatch length and the 1.6th power of the hatch width, i.e. broad hatchways are more expensive than long, narrow ones.

Relationship of unit costs

Unit costs relating to steel weight and machinery may change with time. However, if their ratio remains constant, the result of the calculation will remain unchanged. If, for example, a design calculation for future application assumes the same rates of increase compared with the present for all the costs entered in the calculation, the result will give the same main dimensions as a calculation using only current data.

Annual income and expenditure

The income of cargo ships depends on the amount of cargo and the freight rates. Both should be a function of speed in a free market. At least the interest of the tied-up capital cost of the cargo should be included as a lower estimate for this speed dependence. The issue will be discussed again in Section 3.4 for the effect of speed.
    Expenditure over the lifetime of a ship includes:

1. *Risk costs*
   Risk costs relating to the ship consist mainly of the following insurance premiums:
   • Insurance on hull and associated equipment.
   • Insurance against loss or damage by the sea.
   • Third-party (indemnity) insurance.
   Annual risk costs are typically 0.5% of the production costs.
2. *Repair and maintenance costs*
   The repair and maintenance costs can be determined using operating cost statistics from suitable basis ships, usually available in shipping companies.
3. *Fuel and lubricating costs*
   These costs depend on engine output and operating time.
4. *Crew costs*
   Crew costs include wages and salaries including overtime, catering costs, and social contributions (health insurance, accident and pension insurance, company pensions). Crewing requirements depend on the engine power, but remain unchanged for a wide range of outputs for the same system. Thus crew costs are usually variation-independent. If the optimization result shows a different crewing requirement from the basis ship, crew cost differences can be included in the model and the calculation repeated.
5. *Overhead costs*
   • Port duties, lock duties, pilot charges, towage costs, haulage fees.
   • Overheads for shipping company and broker.
   • Hazard costs for cargo (e.g. insurance, typically 0.2–0.4% of cargo value).
   Port duties, lock duties, pilot charges and towage costs depend on the tonnage. The proportion of overheads and broker fees depend on turnover

and state of employment. All overheads listed here are variation-
independent for constant ship size.

6. *Costs of working stock and extra equipment*
   These costs depend on ship size, size of engine plant, number of crew, etc.
   The variation-dependence is difficult to calculate, but the costs are small
   in relation to other cost types mentioned. For this reason, differences in
   working-stock costs may be neglected.

7. *Cargo-handling costs*
   Cargo-handling costs are affected by ship type and the cargo-handling
   equipment both on board and on land. They are largely variation-
   independent for constant ship size.

Taxes, interest on loans covering the initial building costs and inflation have
only negligible effects on the optimization of main dimensions and can be
ignored.

The 'cost difference' method

Cash flow and initial costs can be optimized by considering only the differences
with respect to the 'basis ship'. This simplifies the calculation as only variation-
dependent items remain. The difference costs often give more reliable figures.

Objective function for initial costs optimization

The initial difference costs consist of the sum of hull steel difference costs and
propulsion unit difference costs:

$$\Delta K_G[\text{MU}] = W_{St_0} \cdot k_{St_0} - W_{St_n} \cdot k_{St_n} + \Delta K_M \cdot C_M$$
$$= W_{St_0} \cdot k_{St_0} - W_{St_n} \cdot k_{St_n} + \Delta P_B \cdot k_M \cdot C_M$$

ΔK_G [MU]	difference costs for the initial costs
W_{St_0} [t]	hull steel weight for basis variant
W_{St_n} [t]	hull steel weight for variant n
k_{St} [MU/t]	specific costs of installed steel
ΔK_M [MU]	difference costs for the main engine
C_M	factor accounting for the difference costs of the 'remaining parts' of the propulsion unit
ΔP_B [kW]	difference in the required propulsion power
k_M [MU/kW]	specific costs of engine power

In some cases the sum of the initial difference costs should be supplemented
further by the equipment difference costs.

Objective function for yield optimization

The yield itself is not required, only the variant which maximizes yield. Again,
only the variation-dependent cash flow needs to be considered. The most
important items are the differences in:

1. Initial costs
2. Fuel and lubricant costs
3. Repair and insurance costs
4. Net income if variation-dependent

The power requirements are a function of trial speed, therefore the initial costs of the propulsion unit depend on the engine requirements under trial speed conditions. The fuel costs should be related to the service speed. The annual fuel and lubricant costs then become:

$$k_{f+l}[\text{MU/yr}] = P_{B,D} \cdot F \cdot (k_f \cdot s_f + k_l \cdot s_l)$$

$P_{B,D}$ [kW]   brake power at service speed
F [h]   annual operating time
k_f [MU/t]   cost of 1 t of fuel (or heavy oil)
s_f [t/kWh] specific fuel consumption
k_l [MU/t]   cost of 1 t of lubricating oil
s_l [t/kWh] specific lubricant consumption

Discontinuities in propulsion unit costs

Standardized propulsion unit elements such as engines, gears, etc. introduce steps in the cost curves (Figs 3.4 and 3.5). The stepped curve can have a minimum on the faired section or at the lower point of a break. With the initial costs, the optimum is always situated at the beginning of the curve to the right of the break. Changing from a smaller to a larger engine reduces the engine loading and thus repair costs. The fuel costs are also stepped where the number of cylinders changes (Fig. 3.6). At one side of the break point the smaller engine is largely fully loaded. On the other side, the engine with one more cylinder has a reduced loading, i.e. lower fuel consumption. Thus when both initial costs and annual costs are considered the discounted cash flow is quasi-continuous.

The assumption of constant speed when propulsion power is changed in steps is only an assumption for comparison when determining the optimum main dimensions. In practice, if the propulsion plant is not fully employed, a higher speed is adopted.

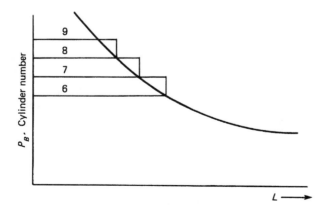

Figure 3.4 Propulsion power P_B and corresponding engine cylinder number as a function of ship's length

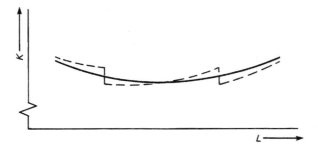

Figure 3.5 Effect of a change in number of engine cylinders on the cost of the ship

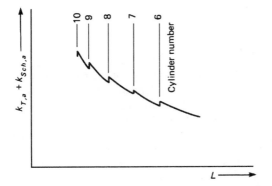

Figure 3.6 Annual fuel and lubricant costs $(k_f + k_l)$ as a function of number of engine cylinders and ship's length

3.4 Discussion of some important parameters

Width

A lower limit for B comes from requiring a minimum metacentric height $\overline{GM}$ and, indirectly, a maximum possible draught. The $\overline{GM}$ requirement is formulated in an inequality requiring a minimum value, but allowing larger values which are frequently obtained for tankers and bulkers.

Length

Suppose the length of a ship is varied while cargo weight, deadweight and hold size, but also $A_M \cdot L$, B/T, B/D and C_B are kept constant (Fig. 3.7). (Constant displacement and underdeck volume approximate constant cargo weight and hold capacity.) Then a 10% increase in length will reduce A_M by 10%. D, B and T are each reduced by around 5%. L/B and L/D are each increased by around 16%.

For this kind of variation, increasing length has these consequences:

1. Increase in required regulation freeboard with decrease in existing freeboard.
2. Decrease in initial stability.

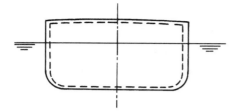

Figure 3.7 Variation of midship section area A_M with proportions unchanged

3. Better course-keeping ability and poorer course-changing ability.
4. Increase in steel weight.
5. Decrease in engine output and weight—irrespective of the range of Froude number.
6. Decrease in fuel consumption over the same operational distance.

Increase in the regulation freeboard

The existing freeboard is decreased, while the required freeboard is increased (Fig. 3.8). These opposing tendencies can easily lead to conflicts. The freeboard regulations never conflict with a shortening of the ship, if C_B is kept constant.

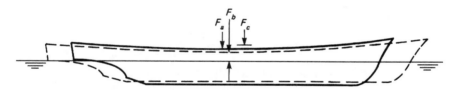

Figure 3.8 Effect of length variation on the freeboard. F_a = freeboard of basis form, F_b = freeboard of distorted ship, F_c = desired freeboard after lengthening

Reduction in initial stability

The optimization often requires constant initial stability to meet the prescribed requirements and maintain comparability. A decrease in $\overline{GM}$ is then, if necessary, compensated by a slight increase of B/T, reducing T and D somewhat. This increases steel weight and decreases power savings.

Course-keeping and course-changing abilities

These characteristics are in inverse ratio to each other. A large rudder area improves both.

Increase in steel weight, decrease in engine output and weight, decrease in fuel consumption

These changes strongly affect the economics of the ship, see Section 3.3.

Block coefficient

Changes in characteristics resulting from reducing C_B:

1. Decrease in regulation freeboard for $C_B < 0.68$ (referred to 85% D).
2. Decrease in area below the righting arm curve if the same initial stability is used.
3. Slight increase in hull steel weight.
4. Decrease in required propulsion power, weight of the engine plant and fuel consumption.
5. Better seakeeping, less added resistance in seaway, less slamming.
6. Less conducive to port operation as parallel middlebody is shorter and flare of ship ends greater.
7. Larger hatches, if the hatch width increases with ship width. Hatch covers therefore are heavier and more expensive. The upper deck area increases.
8. Less favourable hold geometry profiles. Greater flare of sides, fewer rectangular floor spaces.
9. The dimensional limits imposed by slipways, docks and locks are reached earlier.
10. Long derrick and crane booms, if the length of these is determined by the ship's width and not the hatch length.

Initial stability

$\overline{GM}$ remains approximately constant if B/T is kept constant. However, the prescribed $\overline{GM}$ is most effectively maintained by varying the width using Mühlbradt's formula:

$$B = \frac{B_0}{C\left[(C_B/C_{B0})^2 - 1\right] + 1}$$

$C = 0.12$ for passenger and containerships
$C = 0.16$ for dry cargo vessels and tankers.

Seakeeping

A small C_B usually improves seakeeping. Since the power requirement is calculated for trial conditions, no correction for the influence of seastate is included. Accordingly, the optimum C_B for service speed should be somewhat smaller than that for trial speed. There is no sufficiently simple and accurate way to determine the power requirement in a seastate as a function of the main dimensions. Constraints or the inclusion of some kind of consideration of the seakeeping are in the interest of the ship owner. If not specified, the shipyard designer will base his optimization on trial conditions.

Size of hold

For general cargo ships, the required hold size is roughly constant in proportion to underdeck volume. For container and ro-ro ships, reducing C_B increases the 'noxious spaces' and more hold volume is required.

Usually the underdeck volume $\nabla_D = L \cdot B \cdot D \cdot C_{BD}$ is kept constant. Any differences due to camber and sheer are either disregarded or taken as constant over the range of variation. C_{BD} can be determined with reasonable accuracy by empirical equations:

$$C_{BD} = C_B + c \cdot \left(\frac{D}{T} - 1\right) \cdot (1 - C_B)$$

with $c = 0.3$ for U-shaped sections and $c = 0.4$ for V-shaped sections.

With the initial assumption of constant underdeck volume, the change in the required engine room size, and any consequent variations in the unusable spaces at the ship's ends and the volume of the double bottom are all initially disregarded. A change in engine room size can result from changes in propulsion power and in the structure of the inner bottom accommodating the engine seatings.

The effect on cost

A C_B variation changes the hull steel and propulsion system costs. Not only the steel weight, but also the price of the processed ton of steel is variation-dependent. A ton of processed steel of a ship with full C_B is relatively cheaper than that of a vessel with fine C_B.

The specific costs of hull steel differ widely over the extent of the hull. We distinguish roughly the following categories of difficulty:

1. Flat areas with straight sections in the parallel middlebody.
2. Flat areas with straight sections not situated in the parallel middlebody, e.g. a piece of deck without sheer or camber at the ship's ends. More work results from providing an outline contour adapted to the outer shell and because the shortening causes the sections to change cross-section also.
3. Slightly curved areas with straight or curved sections. The plates are shaped locally using forming devices, not pre-bent. The curved sections are pre-formed.
4. Areas with a more pronounced curvature curved only in one direction, e.g. bilge strake in middlebody. The plates are rolled cold.
5. Medium-curved plates curved multidimensionally, e.g. some of those in the vicinity of the propeller aperture. These plates are pressed and rolled in various directions when cold.
6. Highly curved plates curved multidimensionally, e.g. the forward pieces of bulbous bows. These plates are pressed or formed when hot.

Decreasing C_B complicates design and construction, thus increasing costs:

1. More curved plates and sections, fewer flat plates with rectangular boundaries.
2. Greater expenditure on construction details.
3. Greater expenditure on wooden templates, fairing aids, gauges, etc.
4. More scrap.
5. More variety in plates and section with associated costs for storekeeping and management.

An increase in C_B by $\Delta C_B = 0.1$ will usually increase the share of the weight attributable to the flat areas of the hull (group (1) of the above groups) by 3%. About 3% of the overall hull steel will move from groups (3)–(5) to groups (1) and (2). The number of highly curved plates formed multidimensionally (group (6)) is hardly affected by a change in C_B. The change in weight of all curved plates and sections of the hull depends on many factors. It is approximately $0.33 \Delta C_B \cdot$ hull steel weight.

Speed

The speed can be decisive for the economic efficiency of a ship and influences the main dimensions in turn. Since speed specifications are normally part of the shipping company requirements, the shipyard need not give the subject much consideration. Since only the agreement on trial speed, related to smooth water and full draught, provides both shipyard and shipping company with a clear contractual basis, the trial speed will be the normal basis for optimization. However, the service speed could be included in the optimization as an additional condition. If the service speed is to be attained on reduced propulsion power, the trial speed on reduced power will normally also be stated in the contract. Ships with two clearly defined load conditions can have both conditions considered separately, i.e. fully loaded and ballast.

Economic efficiency calculations for the purpose of optimizing speed are difficult to formulate due to many complex boundary conditions. Schedules in a transport chain or food preservation times introduce constraints for speed. (For both fish and bananas, for example, a preservation period of around 17 days is assumed.)

Speed variation may proceed on two possible assumptions:

1. Each ship in the variation series has *constant transportation capacity*, i.e. the faster variant has smaller carrying capacity.
2. Each ship in the variation series has a *constant carrying capacity*, i.e. the faster variant has a greater transportation capacity than the slower one and fewer ships are needed.

Since speed increase with constant carrying capacity increases the transportation capacity, and a constant transportation capacity leads to a change of ship size, it is better to compare the transport costs of 1 t of cargo for various ships on one route than to compare costs of several ships directly.

Essentially there are two situations from which an optimization calculation can proceed:

1. Uncompetitive situation. Here, speed does not affect income, e.g. when producer, shipping company and selling organizations are under the same ownership as in some areas of the banana and oil business.
2. Competitive situation. Higher speed may attract more cargo or justify higher freight rates. This is the prime reason for shipowners wanting faster ships. Both available cargo quantity and freight rate as a functions of speed are difficult to estimate.

In any case, all variants should be burdened with the interest on the tied-up capital of the cargo. For the uncompetitive situation where the shipowner transports his own goods, this case represents the real situation. In the competitive case, it should be a lower limit for attractiveness of the service. If the interest on cargo costs are not included, optimizations for dry cargo vessels usually produce speeds some 2 knots or more below normal.

Closely related with the question of optimum speed is that of port turn-around times. Shortening these by technical or organizational changes can improve the ship's profitability to a greater extent than by optimizing the speed.

Some general factors which encourage higher ship speeds are (Buxton, 1976):

- High-value cargo.
- High freight rates.
- Competition, especially when freight rates are fixed as in Conferences.
- Short turn-around time.
- High interest rates.
- High daily operating costs, e.g. crew.
- Reduced cost of machinery.
- Improved hull form design, reduced power requirements.
- Smoother hulls, both new and in service, e.g. by better coatings.
- Cheap fuel.
- Lower specific fuel consumption.

3.5 Special cases of optimization

Optimization of repeat ships

Conditions for series shipbuilding are different from those for single-ship designs. Some of the advantages of series shipbuilding can also be used in repeat ships. For a ship to be built varying only slightly in size and output from a basis ship, the question arises: 'Should an existing design be modified or a new design developed?' The size can be changed by varying the parallel middlebody. The speed can be changed by changing the propulsion unit. The economic efficiency (e.g. yield) or the initial costs have to be examined for an optimum new design and for modification of an existing design.

The advantages of a repeat design (and even of modified designs where the length of the parallel middlebody is changed) are:

1. Reduced design and detailed construction work can save considerable time, a potentially crucial bargaining point when delivery schedules are tight.
2. Reduced need for jigs for processing complicated components constructed from plates and sections.
3. Greater reliability in estimating speed, deadweight and hold size from a basis ship, allowing smaller margins.
4. Greater accuracy in calculating the initial costs using a 'cost difference' method.

Where no smaller basis ship exists to fit the size of the new design, the objective can still be reached by shortening a larger basis ship. This reduces C_B. It may be necessary to re-define the midship area if more than the length of the parallel middlebody is removed. Deriving a new design from a basis ship of the same speed by varying the parallel middlebody is often preferable to developing a new design. In contrast, transforming a basis ship into a faster ship merely by increasing the propulsion power is economical only within very narrow limits.

Simplified construction of steel hull

Efforts to reduce production costs by simplifying the construction process have given birth to several types of development. The normal procedure employed in cargo shipbuilding is to keep C_B far higher than optimum for resistance. This increases the portion of the most easily manufactured parallel middlebody.

Blohm and Voss adopted a different method of simplifying ship forms. In 1967 they developed and built the *Pioneer* form which, apart from bow and stern bulbs, consisted entirely of flat surfaces. Despite 3–10% lower building costs, increased power requirement and problems with fatigue strength in the structural elements at the knuckles proved this approach to be a dead end.

Another simple construction method commonly used in inland vessels is to build them primarily or entirely with straight frames. With the exception of the parallel middlebody, the outer shell is usually curved only in one direction. This also increases the power requirement considerably.

Ships with low C_B can be simplified in construction—with only little increase in power requirement—by transforming the normally slightly curved surfaces of the outer shell into a series of curved and flat surfaces. The curved surfaces should be made as developable as possible. The flat surfaces can be welded fairly cheaply on panel lines. Also, there is less bending work involved. The difference between this and the *Pioneer* form is that the knuckles are avoided. C_B is lower than in the *Pioneer* class and conventional ships. Optimization calculations for simple forms are more difficult than for normal forms since often little is known about the hydrodynamic characteristics and building costs of simplified ship forms.

There are no special methods to determine the resistance of simplified ships, but CFD methods may bring considerable progress within the next decade. Far more serious is the lack of methods to predict the building costs by consideration of details of construction (Kaeding, 1997).

Optimizing the dimensions of containerships

The width

The effective hold width of containerships corresponds to the hatch width. The area on either side of the hatch which cannot be used for cargo is often used as a wing tank. Naturally, the container stowage coefficient of the hold, i.e. the ratio of the total underdeck container volume to the hold volume, is kept as high as possible. The ratio of container volume to gross hold volume (including wing tanks) is usually 0.50–0.70. These coefficients do not take into account any partial increase in height of the double bottom. The larger ratio value applies to full ships with small side strip width and the smaller to fine vessels and greater side strip widths.

For constant C_B, a high container stowage coefficient can best be attained by keeping the side strip of deck abreast of the hatches as narrow as possible. Typical values for the width of this side strip on containerships are:

For small ships:	$\approx 0.8–1.0\,\mathrm{m}$
For medium-sized ships:	$\approx 1.0–1.5\,\mathrm{m}$
For larger ships:	$\approx 1.2–2.0\,\mathrm{m}$

The calculated width of the deck strip adjacent to the hatches decreases relative to the ship's width with increasing ship size. The variation in the figure also decreases with size.

If the ship's width were to be varied only in steps as a multiple of the container width, the statistics of the containership's width would indicate a stepped or discontinuous relationship. However, the widths are statistically distributed fairly evenly. The widths can be different for a certain container number stowed across the ship width, and ships of roughly the same width may even have a different container number stowed across the ship. The reason is that besides container stowage other design considerations (e.g. stability, carrying capacity, favourable proportions) influence the width of containerships. The difference between the continuous variation of width B and that indicated by the number and size of containers is indicated by the statistically determined variation in the wing tank width, typically around half a container width. The practical compromise between strength and construction considerations on the one hand and the requirement for good utilization on the other hand is apparently within this variation.

The length

The length of containerships depends on the hold lengths. The hold length is a 'stepped' function. However, the length of a containership depends not only on the hold lengths. The length of the fore peak may be varied to achieve the desired ship length. Whether the fore end of the hold is made longer or shorter is of little consequence to the container capacity, since the fore end of the hatch has, usually, smaller width than midships, and the hold width decreases rapidly downwards.

The depth

Similarly the depth of the ship is not closely correlated to the container height, since differences can be made up by the hatchway coaming height. The double bottom height is minimized because wing tanks, often installed to improve torsional rigidity, ensure enough tank space for all purposes.

Optimization of the main dimensions

The procedure is the same as for other ships. Container stowage (and thus hold space not occupied by containers) are included at a late stage of refining the optimization model. This subsequent variation is subject to, for example, stability constraints.

The basis variant is usually selected such that the stowage coefficient is optimized, i.e. the deck strips alongside the hatches are kept as narrow as possible. If the main dimensions of the ship are now varied, given constant underdeck capacity and hold size, the number of containers to be stowed below deck will no longer be constant. So the main dimensions must be corrected. This correction is usually only marginal.

Since in slender ships the maximum hold width can only be fully utilized for a short portion of the length, a reduction in the number of containers to be stowed across the width of the midship section would only slightly decrease the number of containers. So the ratio of container volume to hold volume will change less when the main dimensions are varied on slender containerships than on fuller ships.

3.6 Developments of the 1980s and 1990s

Concept exploration models

Concept exploration models (CEMs) have been proposed as an alternative to 'automatic' optimization. The basic principle of CEMs is that of a direct search optimization: a large set of candidate solutions is generated by varying design variables. Each of these solutions is evaluated and the most promising solution is selected. However, usually all solutions are stored and graphically displayed so that the designer gets a feeling for how certain variables influence the performance of the design. It thus may offer more insight to the design process. However, this approach can quickly become impractical due to efficiency problems. Erikstad (1996) gives the following illustrating example: given ten independent design variables, each to be evaluated at ten different values, the total number of combinations becomes 10^{10}. If we assume that each design evaluation takes 1 millisecond, the total computer time needed will be 10^7 seconds—more than 3 months.

CEM applications have resorted to various techniques to cope with this efficiency problem:

- Early rejection of solutions not complying with basic requirements (Georgescu et al., 1990).
- Multiple steps methods where batches of design variables are investigated serially (Nethercote et al., 1981).
- Reducing the number of design variables (Erikstad, 1994).
- Increasing the step length.

Erikstad (1994) offers the most promising approach, which is also attractive for steepness search optimization. He presents a method to identify the most important variables in a given design problem. From this, the most influential set of variables for a particular problem can be chosen for further exploration in a CEM. The benefit of such a reduction in problem dimension while keeping the focus on the important part of the problem naturally increases rapidly with the dimension of the initial problem. Experience of the designer may serve as a short cut, i.e. select the proper variables without a systematic analysis, as proposed by Erikstad.

Among the applications of CEM for ship design are:

- A CEM for small warship design (Eames and Drummond, 1977) based on six independent variables. Of the 82 944 investigated combinations, 278 were acceptable and the best 18 were fully analysed.
- A CEM for naval SWATH design (Nethercote et al., 1981) based on seven independent variables.
- A CEM for cargoship design (Georgescu et al., 1990; Wijnholst, 1995) based on six independent variables.

CEM incorporating knowledge-based techniques have been proposed by Hees (1992) and Erikstad (1996), who also discuss CEM in more detail.

Optimization shells

Design problems differ from most other problems in that from case to case different quantities are specified or unknown, and the applicable relations may change. This concerns both economic and technical parts of the optimization model. In designing scantlings for example, web height and flange width may be variables to be determined or they may be given if the scantling continues other structural members. There may be upper bounds due to spatial limitations, or lower bounds because crossing stiffeners, air ducts, etc. require a structural member to be a certain height. Cut-outs, varying plate thickness, and other structural details create a multitude of alternatives which have to be handled. Naturally most design problems for whole ships are far more complex than the sketched 'simple' design problem for scantlings.

Design optimization problems require in most cases tailor-made models, but the effort of modifying existing programs is too tedious and complex for designers. This is one of the reasons why optimization in ship design has been largely restricted to academic applications. Here, methods of 'machine intelligence' may help to create a suitable algorithm for each individual design problem. The designer's task is then basically reduced to supplying:

- a list of specified quantities;
- a list of unknowns including upper and lower bounds and desired accuracy;
- the applicable relations (equations and inequalities).

In conventional programming, it is necessary to arrange relations such that the right-hand sides contain only known quantities and the left-hand side only one unknown quantity. This is not necessary in modern optimization shells. The relations may be given in arbitrary order and may be written in the most convenient way, e.g. $\nabla = C_B \cdot L \cdot B \cdot T$, irrespective of which of the variables are unknown and which are given. This 'knowledge base' is flexible in handling diverse problems, yet easy to use.

Such optimization shells include CHWARISMI (Söding, 1977) and DELPHI (Gudenschwager, 1988). These shells work in two steps. In the first step the designer compiles all relevant 'knowledge' in the form of relations. The shell checks if the problem can be solved at all with the given relations and which of the relations are actually needed. Furthermore, the shell checks if the system of relations may be decomposed into several smaller systems which can be solved independently. After this process, the modified problem is converted into a Fortran program, compiled and linked. The second step is then the actual numerical computation using the Fortran program.

The following example illustrates the concept of such an optimization shell. The problem concerns the optimization of a containership and is formulated for the shell in a quasi-Fortran language:

```
      PROGRAM CONT2
C Declaration of variables to be read from file
C TDW      t          deadweight
C VORR     t          provisions
C VDIEN    m/s        service speed
C TEU      -          required TEU capacity
C TUDMIN              share of container capacity underdeck (<1.)
C NHUD                number of bays under deck
C NHOD                number of bays on deck
```

```
C NNUD                number of stacks under deck
C NNOD                number of stacks on deck
C NUEUD               number of tiers under deck
C MDHAUS    t         mass of deckhouse
C ETAD      -         propulsive efficiency
C BMST      t/m**3    weight coefficient for hull
C BMAUE     t/m**2    weight coefficient for E&O
C BMMA      t/kW      weight coefficient for engine
C BCST      DM/t      cost per ton steel hull
C BCAUE     DM/t      cost per ton E&O (initial)
C BCMA      DM/t      cost per ton engine (initial)
C
C Declaration of other variables
C LPP       m         length between perpendiculars
C BREIT     m         width
C TIEF      m         draft
C CB                  block coefficient
C VOL       m**3      displacement volume
C CBD                 block coefficient related to main deck
C DEPTH     m         depth
C LR        m**3      hold volume
C TEUU                number of containers under deck
C TEUO                number of containers on deck
C NUEOD               number of tiers on deck
C GM        m         metacentric height
C PD        kW        delivered power
C MSTAHL    t         weight of steel hull
C MAUE      t         weight of E&O
C MMASCH    t         machinery weight
C CSCHIF    DM        initial cost of ship
C CZUTEU    DM/TEU    initial cost/carrying capacity
C
C Declare type of variables
      REAL BCAUE, BCMA, BCST, BMAUE, BMMA, BMST, ETAD, MDHAUS,
      REAL TEU, TDW, TUDMIN, VDIEN, VORR
      REAL NHOD, NHUD, NNOD, NNUD, NUEUD
C Input from file of required values
      CALL INPUT(BCAUE,BCMA,BCST,BMAUE,BMMA,BMST,ETAD,MDHAUS,
     &          TDW,TEU,TUDMIN,VDIEN,VORR,NHOD,NHUD,NNOD,NNUD,NUEUD)
C      unknowns          start    initial   lower      upper
C                        value    stepsize  limit      limit
      UNKNOWNS LPP    (120.     , 20.0 ,   50.0  ,    150.0),
     &         BREIT  (20.      ,  4.0 ,   10.0  ,     32.2),
     &         TIEF   (5.       ,  2.0 ,    4.0  ,      6.4),
     &         CB     (0.6      ,  0.1 ,    0.4  ,     0.85),
     &         VOL    (7200.    ,500.0 ,1000.0   ,  30000.0),
     &         CBD    (0.66     ,  0.1 ,     .5  ,     0.90),
     &         DEPTH  (11.      ,  2.0 ,    5.0  ,     28.0),
     &         LR     (12000.   ,500.0 ,10000.0  ,  50000.0),
     &         TEUU   (.5*TEU   , 20.0 ,    0.0  ,    TEU  ),
     &         TEUO   (.5*TEU   , 20.0 ,    0.0  ,    TEU  ),
     &         NUEOD  (2.       ,   .1 ,    1.0  ,      4.0),
     &         GM     (1.0      ,  0.1 ,    0.4  ,      2.0),
     &         PD     (3000.    ,100.0 ,  200.0  ,  10000.0),
     &         MSTAHL (1440.    ,100.0 ,  200.0  ,  10000.0),
     &         MAUE   (360.     , 50.0 ,   50.0  ,   2000.0),
```

```
      &              MMASCH(360.      , 50.0 ,  50.0  ,   2000.0),
      &              CSCHIF(60.E6    ,1.E6  , 2.E6   ,    80.E6 ),
      &              CZUTEU(30000.   ,5000. , 10000. ,   150000.)
C  ****   Relations decribing the problem  ****
C mass and displacement
      VOL      =  LPP*BREIT*TIEF*CB
      VOL*1.03 =  MSTAHL + MDHAUS + MAUE + MMASCH + TDW
      MSTAHL   =  STARUM (BMST,LPP,BREIT,TIEF,DEPTH,CBD)
      MAUE     =  BMAUE*LPP*BREIT
      MMASCH   =  BMMA*(PD/0.85)**0.89
C stability
      GM       =  0.43*BREIT - ( MSTAHL*0.6*DEPTH
      &                         +MDHAUS*(DEPTH+6.0)
      &                         +MAUE*1.05*DEPTH
      &                         +MMASCH*0.5*DEPTH
      &                         +VORR*0.4*DEPTH
      &                         +TEUU*MCONT*(0.743-0.188*CB)
      &                         +TEUO*MCONT*(DEPTH+2.1+0.5*NUEOD*HCONT)
      &                         )/VOL/1.03
C  hold
      CBD      =  CB+0.3*(DEPTH-TIEF)/TIEF*(1.-CB)
      LR       =  LPP*BREIT*DEPTH*CBD*0.75
C container stowing / main dimensions
      LPP   .GE. (0.03786+0.0016/CB**5)*LPP
      &            +0.747*PD**0.385
      &            +NHUD*(LCONT+1.0)
      &            +0.07*LPP
      LPP   .GE. 0.126*LPP+13.8
      &            +(NHOD-2.)*(LCONT+1.0)
      &            +0.07*LPP
      BREIT .GE. 2.*2.0+BCONT*NNUD+(NNUD+1.)*0.25
      BREIT .GE. 0.4 + BCONT*NNOD+(NNOD-1)*0.04
      DEPTH .GE. (350+45*BREIT)/1000. + NUEUD*HCONT - 1.5
      TEU  = TEUU +TEUO
      TEUU .GE. TUDMIN*TEU
      TEUU = (0.9*CB+0.26)*NHUD*NNUD*NUEUD
      TEUO = (0.5*CB+0.55)*NHOD*NNOD*NUEOD
C  propulsion
      PD       =  VOL**0.567*VDIEN**3.6 / (153.*ETAD)
C  building cost
      CSCHIF   =  BCST*MSTAHL*SQRT(.7/CB)+ BCAUE*MAUE + BCMA*MMASCH
      CZUTEU   =  CSCHIF/(TEUU+TEUO)
C  freeboard approximation
      DEPTH - TIEF . GE. 0.025*LPP
C  L/D ratio
      LPP/DEPTH.GE.8.
      LPP/DEPTH.LE.14.
C Criterion: minimize initial cost/carried container
      MINIMIZE CZUTEU
      SOLVE
C Output
      CALL OUTPUT(LPP,BREIT,TIEF,CB,VOL,CBD,DEPTH,LR,TEUU,TEUO,NUEOD,
      &              GM,PD,MSTAHL,MAUE,MMASCH,CSCHIF,CZUTEU)
      END

      REAL FUNCTION STARUM(BMST,LPP,B,T,D,CBD)
```

```
C   weight of steel hull following SCHNEEKLUTH, 1985
        REAL B, BMST, CBD, C1, D, LPP, T, VOLU
        VOLU=LPP*B*D*CBD
        C1=BMST*(1.+0.2E-5*(LPP-120.)**2)
        STARUM=VOLU*C1
    &        *(1.+0.057*(MAX(10.,LPP/D)-12.))
    &        *SQRT(30./(D+14.))
    &        *(1.+0.1*(B/D-2.1)**2)
    &        *(1.+0.2*(0.85-T/D))
    &        *(0.92+(1.-CBD)**2)
        END
```

The example shows that the actual formulation of the problem is relatively easy, especially since it can be based on existing Fortran procedures (steel weight in this example).

Even an optimization shell is not foolproof and errors occur frequently when beginners start using the shell. Not the least of the problems is that users formulate problems which allow no solution as improper constraints are imposed.

Another problem is that, in reality, many design problems are not so clearly defined. While there are, in principle, techniques to include uncertainty in the optimization (other than through sensitivity analyses) (e.g. Schmidt, 1996), extended functionality always comes at the price of added complexity for the user, which in our experience at present prevents acceptance.

Optimization shells of the future should try to extend functionality without sacrificing user-friendliness. Perhaps further incorporation of knowledge-based techniques, namely in formulating and interpreting results, could be the path to a solution. But even the most 'intelligent' system will not relieve the designer of the task to think and to decide.

3.7 References

BENFORD, H. (1965). Fundamentals of ship design economics. Department of Naval Architects and Marine Engineers, Lecture Notes, University of Michigan

BUXTON, I. L. (1976). Engineering economics and ship design. British Ship Research Association report, 2nd edn

EAMES, M. C. and DRUMMOND, T. G. (1977). Concept exploration—an approach to small warship design. *Trans. RINA* **119**, p. 29

ERIKSTAD, S. O. (1994). Improving concept exploration in the early stages of the ship design process. *5th International Marine Design Conference*, Delft, p. 491

ERIKSTAD, S. O. (1996). A Decision Support Model for Preliminary Ship Design. Ph.D. thesis, University of Trondheim

GEORGESCU, C., VERBAAS, F. and BOONSTRA, H. (1990). Concept exploration models for merchant ships. *CFD and CAD in Ship Design*, Elsevier Science Publishers, p. 49

GUDENSCHWAGER, H. (1988). Optimierungscompiler und Formberechnungsverfahren: Entwicklung und Anwendung im Vorentwurf von RO/RO-Schiffen. IfS-Report 482, University of Hamburg

HEES, M. VAN (1992). Quaestor: A knowledge-based system for computations in preliminary ship design. PRADS' 92, NewCastle, p. 21284

JANSON, C. E. (1997). Potential Flow Panel Methods for the Calculation of Free-surface Flows with Lift. Ph.D. thesis, Gothenborg

KAEDING, P. (1997). Ein Ansatz zum Abgleich von Fertigungs- und Widerstandsaspekten beim Formentwurf. *Jahrbuch Schiffbautechn. Gesellschaft*

KEANE, A. J., PRICE, W. G. and SCHACHTER, R. D. (1991). Optimization techniques in ship concept design. *Trans. RINA* **133**, p. 123

KERLEN, H. (1985). Über den Einfluß der Völligkeit auf die Rumpfstahlkosten von Frachtschiffen. IfS Rep. 456, University of Hamburg

LIU, D., HUGHES, O. and MAHOWALD, J. (1981). Applications of a computer-aided, optimal preliminary ship structural design method. *Trans. SNAME* **89**, p. 275

MALONE, J. A., LITTLE, D. E. and ALLMAN, M. (1980). Effects of hull foulants and cleaning/coating practices on ship performance and economics. *Trans. SNAME* **88**, p. 75

MALZAHN, H., SCHNEEKLUTH, H. and KERLEN, H. (1978). OPTIMA, Ein EDV-Programm für Probleme des Vorentwurfs von Frachtschiffen. Report 81, Forschungszentrum des Deutschen Schiffbaus, Hamburg

NETHERCOTE, W. C. E., ENG, P. and SCHMITKE, R. T. (1981). A concept exploration model for SWATH ships. *The Naval Architect*, p. 113

PAPANIKOLAOU, A. and KARIAMBAS, E. (1994). Optimization of the preliminary design and cost evaluation of fishing vessel. *Schiffstechnik* **41**, p. 46

RAY, T. and SHA, O .P. (1994). Multicriteria optimization model for containership design. *Marine Technology* **31/4**, p. 258

SCHMIDT, D. (1996). Programm-Generatoren für Optimierung unter Berücksichtigung von Unsicherheiten in schiffstechnischen Berechnungen. IfS Rep. 567, University of Hamburg

SCHNEEKLUTH, H. (1957). Die wirtschaftliche Länge von Seefrachtschiffen und ihre Einfluß faktoren, *Schiffstechnik* **13**, p. 576

SCHNEEKLUTH, H. (1967). Die Bestimmung von Schiffslänge und Blockkoeffizienten nach Kostengesichtspunkten, *Hansa*, p. 367

SEN, P. (1992). Marine design: The multiple criteria approach. *Trans. RINA*, p. 261

SÖDING, H. (1977). Ship design and construction programs (2). *New Ships* **22/8**, p. 272

TOWNSIN, R. L., BYRNE, D., SVENSEN, T. E. and MILNE, A. (1981). Estimating the technical and economic penalties of hull and propeller roughness. *Trans. SNAME* **89**, p. 295

WIJNHOLST, N. (1995). *Design Innovation in Shipping*. Delft University Press

WINKLE, I. E. and BAIRD, D. (1985). Towards more effective structural design through synthesis and optimisation of relative fabrication costs. *Naval Architect*, p. 313; also in *Trans. RINA* (1986), p. 313

4

Some unconventional propulsion arrangements

4.1 Rudder propeller

Rudder propellers (slewable screw propellers) (Bussemaker, 1969)—with or without nozzles—are not just a derivative of the well-known outboarders for small boats. Outboarders can only slew the propeller by a limited angle to both sides, while rudder propellers can cover the full 360°. Slewing the propeller by 180° allows reversal of the thrust. This astern operation is much more efficient than for conventional propellers turning in the reverse direction. By 1998, rudder propellers were available at ratings up to 4000 kW.

4.2 Overlapping propellers

Where two propellers are fitted, these can be made to overlap (Pien and Strom-Tejsen, 1967; Munk and Prohaska, 1968) (Fig. 4.1). As early as the 1880s, torpedo boats were fitted with overlapping propellers by M. Normand at the French shipyard. The propellers turned in the same direction partially regaining the rotational energy. Model tests in Germany in the 1970s covered only cases for oppositely turning propellers. Better results were obtained for propellers which turned outside on the topside.

Overlapping propellers have rarely been used in practice, although the theory has been extensively investigated in model tests. It differs from conventional arrangements in the following ways:

1. The total jet area is smaller—this reduces the ideal efficiency.
2. The propellers operate in an area of concentrated wake. This increases hull efficiency $\eta_H = (1 - t)/(1 - w)$.
3. There may be some effects from mutual interaction.
4. Parallel shafts with a small axial separation provide less propeller support. Propeller support is improved if the smaller propeller separation is used with a rearwards converging shaft arrangement. This also makes engine arrangement easier.
5. Recovery of rotational energy with both propellers turning in the same direction.
6. The resistance of open-shaft brackets and shafts placed obliquely in the flow is lower than in the conventional twin-screw arrangement.

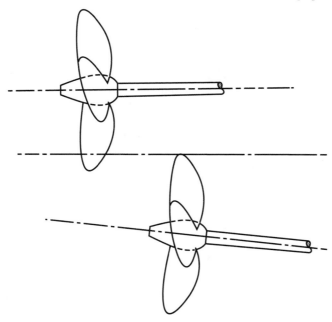

Figure 4.1 Overlapping propellers may be designed with converging shafts as shown, or parallel shafts

The decrease in jet area and the possibility of utilizing the concentrated wake mutually influence efficiency. The overall propulsion efficiency attained is higher than that using a conventional arrangement. The resistance of the struts and shafts is reduced by around one-third with subsequent reductions in required power.

Overlapping propellers with aft slightly converging shafts feature two advantages:

+ Engine arrangement is easier.
+ The course-changing ability is increased.

The convergence of the shafts leads to a strong rudder moment if only one of the propellers is working. Therefore it should be determined in model tests whether the ship is able to steer straight ahead if one of the propulsion systems fails. Such a check is highly recommended for convergence angles (towards the centreplane) of 3° or more.

Interaction effects can cause vibration and cavitation. Both can be overcome by setting the blades appropriately. The port and starboard propellers should have a different number of blades.

The following quantities influence the design:

1. Direction of rotation of the propeller.
2. Distance between shafts.
3. Clearance in the longitudinal direction.
4. Stern shape.
5. Block coefficient.

The optimum direction of rotation with regard to efficiency is top outwards. The flow is then better at the counter and has less tendency to separate. Sometimes an arrangement with both shafts turning in the same direction may be better owing to energy recovery.

The optimum distance between the shafts is 60–80% of the propeller diameter (measured on a containership). The separation in the longitudinal direction has only a slight effect on efficiency and affects primarily the level of vibration.

The U-shaped transverse section, used in single-screw vessels, particularly favours this propeller arrangement—unlike the V form usually found on twin-screw vessels. The overlapping propeller arrangement has more advantages for fuller hull forms, since the possibilities for recovering wake energy are greater. Some of the advantages gained in using overlapping propellers can also be attained by arranging the propellers symmetrically with a small distance between the shafts. With overlapping propellers a single rudder can be arranged in the propeller stream.

4.3 Contra-rotating propellers

Rotational exit losses amount to about 8–10% in typical cargo ships (van Manen and Sentic, 1956). Coaxial contra-rotating propellers (Fig. 4.2) can partially compensate these losses increasing efficiency by up to 6% (Isay, 1964; Lindgren *et al.*, 1968; Savikurki, 1988). To avoid problems with cavitation, the after-propeller should have a smaller diameter than the forward propeller.

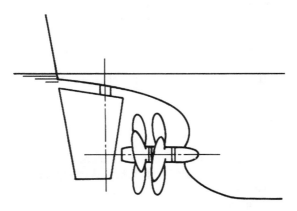

Figure 4.2 Contra-rotating coaxial propellers

Contra-rotating propellers have the following advantages and disadvantages:

+ The propeller-induced heeling moment is compensated (this is negligible for larger ships).
+ More power can be transmitted for a given propeller radius.
+ The propeller efficiency is usually increased.
− The mechanical installation of coaxial contra-rotating shafts is complicated, expensive and requires more maintenance.

− The hydrodynamic gains are partially compensated by mechanical losses in shafting.

Contra-rotating propellers are used on torpedos due to the natural torque compensation. They are also found in some motorboats. For normal ships, the task of boring out the outer shafts and the problems of mounting the inner shaft bearings are not usually considered to be justified by the increase in efficiency, although in the early 1990s some large tankers were equipped with contra-rotating propellers (N. N., 1993; Paetow et al., 1995).

The Grim wheel, Section 4.6, is related to the contra-rotating propeller, but the 'aft' propeller is not driven by a shaft. Unlike a contra-rotating propeller, the Grim wheel turns in the same direction as the propeller.

4.4 Controllable-pitch propellers

Controllable-pitch propellers (CPP) are often used in practice. They feature the following advantages and disadvantages:

+ Fast stop manoeuvres are possible.
+ The main engine does not need to be reversible.
+ CPPs allow the main generator to be driven from the main engine which is efficient and cheap. Thus electricity can be generated with the efficiency of the main engine and using heavy fuel. Variable ship speeds can be obtained with constant propeller rpm as required by the generator.
− Fuel consumption is higher. The higher propeller rpm at lower speed is hydrodynamically suboptimal. CPPs require a thicker hub (0.3–$0.32D$). The pitch distribution is suboptimal. The usual almost constant pitch in the radial direction causes negative angles of attack at the outer radii at reduced pitch, thus slowing the ship down. Therefore CPPs usually have higher pitch at the outer radii and lower pitch at the inner radii. The higher pitch at the outer radii necessitates a larger propeller clearance.
− Higher costs for propeller.

The blades are mounted in either pivot or disc bearings. The pitch-control mechanism is usually controlled by oil pressure or, more rarely, pneumatically. CPPs may have three, four or five blades.

4.5 Kort nozzles

Operating mode

The Kort nozzle is a fixed annular forward-extending duct around the propeller. The propeller operates with a small gap between blade tips and nozzle internal wall, roughly at the narrowest point. The nozzle ring has a cross-section shaped as a hydrofoil or similar section. The basic principle underlying nozzle operation is most simply explained according to Horn (1940) by applying simple momentum theory to the basic law of propulsion. This postulates that, for generation of thrust with good efficiency, the water quantity involved must be as large as possible and the additional velocity imparted thereto must be as small as possible. If, through correct shaping, e.g. provision of an appropriately large inlet opening, propeller operation in the nozzle can be successfully supplied with a larger water quantity than that available to a free propeller of equal diameter at the same thrust, propeller operating conditions are improved (Fig. 4.3). Thrust is additionally generated by the nozzle itself. Due to the larger water quantity, the addition of velocity necessary for thrust generation proves to be smaller. Ideal efficiency rises.

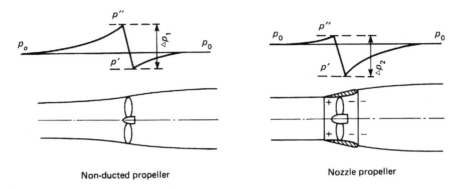

Figure 4.3 Pressure process and flow contraction at a nozzle propeller compared to a free propeller

At equal propeller diameter, a higher inflow velocity at the propeller location is necessarily associated with the increased flowrate. An area of reduced pressure forward of the nozzle propeller, which is more pronounced than that of the free propeller, results from this excess velocity.

The pressure change in the propeller associated with flow acceleration is—at equal thrust—somewhat reduced due to the greater flowrate:

$$\Delta p_2 < \Delta p_1$$

The pressure change is, however, simultaneously displaced by the reduced pressure resulting from the excess velocity at the nozzle inlet to a lower pressure level and thereby its major effect is at the forward nozzle entry. In conjunction with shaping of the nozzle internal wall, this pressure difference dislocation generates a strong underpressure forward of the propeller. Behind the propeller, a weaker, but thrust-generating, overpressure domain occurs in any case where the propeller is arranged at the narrowest point of the nozzle, and this further extends aft to some degree. This generates a negative thrust deduction, equivalent to effective nozzle thrust T_d, which relieves the propeller of part of the total thrust T_0 to be applied.

If a transition is now made from simplified momentum theory to the real propeller, its reduced thrust-loading coefficient

$$C_s = \frac{T}{\rho/2 \cdot V_A^2 \cdot D^2}$$

is substantially changed. At a total thrust T_0, which corresponds to that of the free propeller, the actual propeller thrust T is reduced by the proportion of nozzle thrust T_d as:

$$T = T_0 - T_d$$

The inflow velocity V_A relative to the free propeller is increased. A higher propeller efficiency η_0 results from the significantly reduced thrust-loading coefficient, i.e. at equivalent total system (nozzle plus propeller) thrust, lower propulsive power P_D is required relative to the free propeller. The higher efficiency is also expressed in the reduced, or even completely suppressed, flow contraction associated with the magnitude of velocity change. These positive effects—at least at higher load factors—largely outweigh the additional specific resistance of the nozzle itself.

In accordance with the extensive nozzle effect theory enunciated by Horn, the nozzle is treated as an annular foil, which is replaced by a vortex ring on an annular vortex surface and thereby made amenable to calculation (Fig. 4.4).

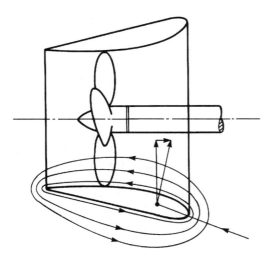

Figure 4.4 Nozzle as foil ring. Section inflow direction, circulating flow, and lift force, together with components directed forward, resulting from propeller operation

The inflow conditions of this foil ring are decisively affected by the propeller incident flow, which is at an angle to the shaft. The section thus experiences a resultant oblique inflow leading to a circulation flow around the section and a resultant section lifting force. Because of the shape of the nozzle cross-section, this resultant force has a forward-directed component corresponding to nozzle thrust. The nozzle thrust, defined in this approach as the forward component of hydrofoil lifting force, is identical with the resultant force from the previously explained underpressure field.

The increased flowrate, or increased flowrate velocity through the propeller, is now explained on the basis of the circulation flow which, owing to the foil effect, is superimposed on the incident flow of the free propeller. According to this theoretical interpretation, which has become most widespread, Kort nozzles are foil rings that shroud the propeller. Propeller and nozzle ring thereby form a functional unit in which they interact.

Nozzle advantages and disadvantages

+ At high thrust-loading coefficients, better efficiency is obtainable. For tugs and pusher boats, efficiency improvements of around 20% are frequently achievable. Bollard pull can be raised by more than 30%.
+ The reduction of propeller efficiency in a seaway is lower for nozzle propellers than for non-ducted propellers.
+ Course stability is substantially improved by the nozzle.
+ In 'steerable nozzle' versions, the nozzle replaces the rudder. The hull waterlines at nozzle height can be run further aft and thus the waterline endings can be made finer and ship resistance reduced. The steerable nozzle, however, has a somewhat lower efficiency than the fixed nozzle, since the gap between propeller blade tips and nozzle internal wall must be kept slightly larger. There is also less space for the propeller diameter, since the steerable nozzle, unlike conventional fixed nozzles, cannot fit into the stern counter.
− Course-changing ability during astern operation is somewhat impaired.
− Owing to circulation in shallow water, the nozzle propeller tends to draw into itself shingle and stones. Also possible is damage due to operation in ice. This explains the infrequent application on seagoing ships.
− Due to the pressure drop in the nozzle, cavitation occurs earlier.

Kort nozzle history

In 1924, Ludwig Kort (1888–1958) submitted a patent application for a ship fitted with an internal propeller in a tunnel. The bow wave was to be reduced by this flow through the tunnel, though the high additional frictional resistance of the tube had the effect of increasing resistance (Fig. 4.5). In the course of time, the long tube traversing the ship has been compressed into a nozzle ring located outside of the ship. After years of successful engineering work, Kort empirically developed the nozzle, which soon found widespread applications in inland navigation. In 1940, a fundamental theoretical paper addressing the nozzle's mode of operating was published by Horn. Building on these ideas, Amtsberg developed the first nozzle design procedure. In subsequent years, the nozzle form has developed along the foil ring route.

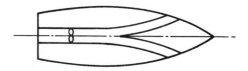

Figure 4.5 Principle of the original Kort nozzle concept

Calculation

Nozzle application criterion

The following criterion, derived from the data of Amtsberg, may be applied as a first approximation to test whether a Kort nozzle offers savings in power output:

$$\frac{P_D}{D^2 \cdot V_A^3} > 1.6$$

P_D [kW]   shaft output,
D [m]   propeller diameter, and
V_A [m/s] inflow velocity of propeller without nozzle.

The following conditions apply:

1. Sectors for nozzle mounting above and a flattening below together come to around 90°.
2. No efficiency loss due to cavitation.
3. The propeller diameter is not restricted by the nozzle.
4. Suitable dimensions for nozzle length, dihedral angle, and profile are selected.

Amtsberg's calculation procedure

The calculation procedure of Amtsberg (1950), see also Horn (1950), reverts to the method proposed by Horn to calculate the nozzle system semi-empirically. In terms of propeller circulation theory, the lifting effect of a foil surface—on the basis of the Kutta and Joukowski hypothesis—may be replaced by a 'line vortex'. The nozzle is then represented by a vortex ring which accelerates the flow in the nozzle. An additional velocity is superposed on the nozzle inflow velocity. Thus, the nozzle generates a negative wake, whose magnitude is determined by the nozzle profile and is numerically determinable using the vortex ring. The major problem centres on the correct determination of nozzle wake factor w_d and nozzle thrust-deduction factor t_D. The inflow velocity to the nozzle (to be determined like the inflow velocity of a non-ducted propeller) differs from that of the propeller in the nozzle:

$$V_A = V \cdot (1 - w) \cdot (1 - w_d)$$

The advance coefficient of the nozzle propeller is

$$J = \frac{V_A}{n \cdot D} = \frac{V \cdot (1 - w) \cdot (1 - w_d)}{n \cdot D}$$

Since the resultant inflow force of the profile is directed inwards and obliquely forward, the nozzle itself has a negative thrust-deduction factor t_D which can also be determined by the procedure. The thrust-deduction factor of the ship is also modified by a nozzle. The 'corrected thrust-deduction factor' of the ship is:

$$t' = t \cdot \tau \sqrt{\frac{1 + C_{Th}}{1 + \tau \cdot C_{Th}}} \quad \text{with} \quad \tau = \frac{1}{1 - t_D}$$

The load ratio τ indicates the proportion of propeller thrust in the total thrust. Amtsberg determined the nozzle wake, nozzle thrust-deduction, and nozzle resistance values needed for a performance calculation for all dimension and loading conditions occurring in practice and presented them non-dimensionally. The procedure was initially based on fully-annular nozzles with NACA profile 4415.

The procedure allows the determination of output requirements and rate of revolution as a function of given ship conditions and nozzle system characteristics. Nozzle system characteristics include those of both propeller and nozzle. Special nozzle characteristics can be optimized by Amtsberg's procedure. Principal characteristics are:

D_I inside diameter ⎫

L nozzle length ⎬ Allowing to optimize the quasi-propulsive coefficient.

α dihedral angle ⎭

The nozzle dihedral angle is the angle between nozzle axis and the line joining the leading and trailing edges of the profile. On the profiles investigated by Amtsberg, an effective angle of attack of 4° is given at a dihedral angle of 0°.

The calculation procedure is:

1. Determination of input values:
   (a) Thrust
   (b) Propeller inflow velocity—without nozzle.
2. Determination of following values included in further calculation:
   (a) Corrected ship thrust-deduction factor.
   (b) Nozzle thrust-deduction factor (relating to a thrust deduction in the ship direction, acting as a positive thrust force).
   (c) Load ratio (indicates propeller thrust proportion).
   (d) Total thrust-loading coefficient of the system (nozzle + propeller).
   (e) Nozzle wake fraction.
   (f) Corrected thrust-loading coefficient of nozzle propeller.

For the calculation, the presentation in Henschke (1965) is simpler and clearer than the original publications.

Advantages of the Amtsberg procedure are:

1. A preliminary investigation can establish whether a nozzle is generally worthwhile.
2. The procedure is widely applicable.

The disadvantages of the procedure may be overcome through minor propeller and nozzle form modifications. Their effects on thrust, efficiency, and rotational speed should be considered through minor corrections. Modifications of the procedure are necessary for:

1. Kaplan propellers, known to offer the best efficiency in tubes (Fig. 4.6).
2. Other nozzle profiles, e.g. for simple-form profiles, with lower initial costs.
3. For rounded trailing edges, which give better astern thrust qualities with minor impairment of ahead thrust (Fig. 4.7).
4. For curvature of the mean camber line to prevent the profile outlet angle from being too small. An excessively small outlet angle means cross-sectional narrowing and thereby larger outlet losses. For a flow cross-section converging aft the pressure also exerts negative thrust on the nozzle internal wall, thus generating a braking force (Fig. 4.8).

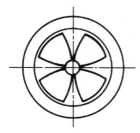

Figure 4.6 Kaplan propeller in a nozzle

Figure 4.7 Nozzle section with sharp and round trailing edge

Deviations from the standard nozzle and standard propeller require some experience in estimating the influence on rotational speed.

Systematic nozzle tests

The published systematic nozzle tests allow simple and reliable calculation of nozzle principal data and also facilitate optimization. Some consideration is given to Kaplan propellers. The structurally simpler Shushkin nozzle forms are to be assessed as though they were standard faired nozzles (as first approximation). Their efficiency is only 1–2% below that of faired nozzles.

Some nozzle characteristics

Some data relating to the magnitude of thrust obtainable with good nozzles are specified below. For pusher boats, the following ahead bollard thrusts are achievable:

For non-ducted propellers	80 N/kW
For propellers in nozzles	100 N/kW

For astern thrust, the following values are achievable:

For non-ducted propellers	60–70 N/kW
For propellers in nozzles	70–75 N/kW

Astern thrust as percentages of ahead thrust are:

For non-ducted propellers	73–82%
For propellers in nozzles	68–77%

These values assume that astern operating or astern thrust properties are considered during nozzle design. If this is not done and, for example, the nozzle trailing edge is kept sharp to optimize forward operating performance, the ratio of astern thrust to ahead thrust amounts to only about 60%.

122

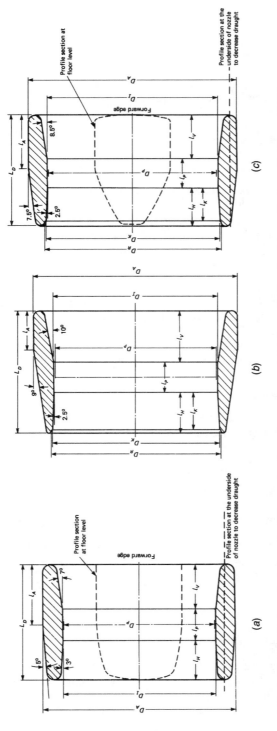

Figure 4.8 Simplified nozzle design: Shushkin nozzles for pushers and conventional tugs (further development Professor Dr Heuser, VBD):

(a): $L_D/D_P = 0.75$; $D_I/D_P = 1.015$; limits: 20 mm $< (D_I - D_P) < 60$ mm; $D_A/D_I = 1.25$; $l_A/L_D = 1.25$; $l_A/L_D = 0.53$; $l_P/L_D = 0.27$, $l_V/L_D = 0.40$; $l_H/L_D = 0.33$
Separation knuckle at front and back depending on specifications
Rounding of nozzle profile at front and back: circular arc

(b): $L_D/D_P = 0.75$; $D_I/D_P = 1.015$; limits: 20 mm $< (D_I - D_P) < 60$ mm; $D_A/D_I = 1.25$; $D_K/D_I = 1.02$; $D_R/D_I = 1.035$; $l_A/L_D = 0.32$; $l_P/L_D = 0.25$, $l_V/L_D = 0.425$; $l_H/L_D = 0.325$; $l_K/l_H = 0.925$

(c): $L_D/D_P = 0.75$; $D_I/D_P = 1.015$; limits: 20 mm $< (D_I - D_P) < 60$ mm; $D_A/D_I = 1.20$; $D_K/D_I = 1.015$; $D_R/D_I = 1.030$ $l_A/L_D = 0.50$; $l_P/L_D = 0.50$, $l_V/L_D = 0.40$; $l_H/L_D = 0.35$; $l_K/l_H = 0.880$
Rounding of nozzle profile at front and back: circular arc

The lower percentage of astern thrust related to ahead thrust for propellers in nozzles compared with propellers without nozzles is due to the fact that, in relation to a non-ducted propeller, ahead thrust with the nozzle can be more substantially improved than astern thrust. Thus, thrust for propellers in nozzles is, in absolute terms, in both ahead and astern directions, greater than for a non-ducted propeller of equal output. For an astern operating fixed-pitch propeller without nozzle, rotational speed falls faster than in the nozzle propeller case, thus again making the propeller with nozzle better than the non-ducted propeller.

Design hints

An improvement in the hydrodynamic performance must be demonstrated to justify the application of Kort nozzles. In a seaway the efficiency of a propeller with nozzle is less reduced than for a non-ducted propeller due to the more axial inflow. The nozzle efficiency increases in a seaway due to the increased thrust-loading coefficient. In total, the nozzle thus decreases the efficiency losses.

When considering if it is worthwhile to install a nozzle, nozzle construction and initial costs play a major role. For performance improvements greater than 7% and propulsive outputs greater than 1000 kW, nozzle acquisition costs are thought to be already lower than the improved propulsive output when considering costs of shaft, exhaust-gas device, etc.

If the installation of Kort nozzles has been decided, nozzle form and arrangement type must be established. For this purpose, the following aspects have to be individually determined:

1. Fixed nozzle or steerable nozzle.
2. Mounting of nozzle by supports or nozzle ring penetration of ship hull.
3. Propeller diameter and nozzle internal diameter.
4. Nozzle profile shape:
   (a) Faired or developable simple-form profile.
   (b) Nozzle aft end sharp or heavily rounded.
   (c) Concentric nozzle or Y-nozzle.
5. Profile length.
6. Nozzle dihedral angle.
7. Special devices for deflection of inflowing objects.
8. Cavitation and air entrainment hazards.
9. Nozzle axis direction.
10. Standard or Kaplan propeller.

These aspects and alternatives are discussed below; see also Philipp *et al.* (1993).

(1) Fixed nozzle or steerable nozzle

Steerable nozzles produce virtually the same rudder effect as a downstream rudder of equal lateral projected area. Since the centre of pressure is located at around one-quarter profile length, with the axis of rotation being arranged at around half profile length to avoid propeller impact against the nozzle internal wall, steerable nozzles are overbalanced. Thus, at small deflection a moment arises acting to increase the deflection.

A rudder-like control surface is therefore frequently suspended behind the propeller on the steerable nozzle to 'balance' the entire system. Thus, at small rudder angle, a net restoring moment occurs. Rudder effect is also increased. A further effect is a partial straightening of the propeller slipstream and an associated enhancement of the quasi-propulsive coefficient. In respect of power saving, steerable nozzles offer advantages and disadvantages:

+ The propeller blade tip circle positioned near the after perpendicular is located further aft than in conventional arrangements. Thus either the horizontal clearance between propeller and stern frame is greater than normal (lower thrust-deduction factor) or the waterlines forward of the propeller have a finer run. Separation resistance may be reduced.
− The clearance between propeller blade tip and nozzle internal wall must be kept 50% larger than for fixed nozzles to avoid blade tip impact. Thus, to rotate the nozzle, a greater lateral distance is required and, due to bearing play, greater vertical distance is also needed. Efficiency drops with gap size.
− Steerable nozzle and propeller diameters, depending on the configuration, are smaller than for fixed nozzles. Steerable nozzles are mostly used on small ships.

(2) Mounting of nozzle by supports or nozzle ring penetration of ship hull

There are various ways to mount Kort nozzles on the hull:

• Steerable nozzles require a cantilever in the plane of the propeller tip.
• There are various options for fixed nozzles: strut construction between nozzle and hull, either by several shaped struts or a flat strut between nozzle and hull.
• Nozzle penetrates hull.

Hull-penetrating nozzles allow the maximum propeller diameter with highest propeller efficiency, but at the price of a 'lost upper sector'. In this sector the nozzle effect is reduced, but not completely lost. The combined propeller efficiency and nozzle efficiency is often optimized when a penetrating nozzle is chosen. The penetrating nozzle also captures more wake and thus improves the hull efficiency. The penetration of the nozzle should be limited such that the inner contour of the nozzle still accelerates the flow, thus reducing the load at the propeller tip (Fig. 4.9). The wedge-shaped gap between counter and outer nozzle contour should be filled by a connecting piece for strength and hydrodynamic reasons. This connecting piece should either taper out or form a connection to the rudder stock.

Steerable nozzles are usually mounted on the rudder stock. For shallow ship sterns and tunnel sterns v.d.Stein has found that it is often better to integrate the nozzle in a rotating plate (Fig. 4.10).

Other structural measures aiding incident flow homogenization are 'skirts' or other control surfaces. Application of skewback propellers may also be appropriate in this context.

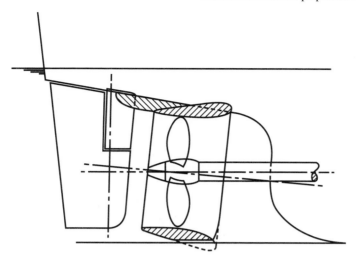

Figure 4.9 Kort nozzle penetrating the hull with a connecting piece for static and hydrodynamic reasons

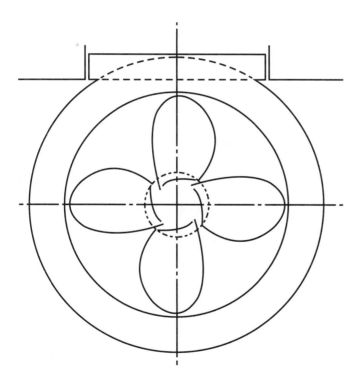

Figure 4.10 Kort nozzle integrated in a rotating plate, offering all the advantages of a hull-penetrating nozzle

(3) Propeller diameter and nozzle internal diameter

Large propeller and nozzle diameters are normally sought. A large propeller diameter restricts other efficiency-enhancing options, e.g.:

1. Nozzle length for pre-selected profile form.
2. Nozzle dihedral angle.

Both factors are still to be discussed. The gap—the difference between nozzle internal radius and propeller radius—should not exceed 0.75% of the radius.

(4) Nozzle profile shape

(a) Faired profiles—simple forms.   For the nozzle profile shape, either faired profiles, e.g. NACA 4415, or simple forms as recommended by Shushkin are used (Fig. 4.8). The simple forms consist of round steel or pipes which at their ends have fully developable surfaces which are essentially conical and cylindrical pieces.

Unlike faired profiles with comparable characteristics, the developable forms are subject to efficiency losses of only 1–2%. Developable forms are frequently used in German inland vessels.

(b) Nozzle after end, sharp or rounded.   As with propeller profiles, the nozzle after end is more heavily rounded if greater value is placed on stopping behaviour. By rounding the nozzle profile end, ahead efficiency falls somewhat. Depending on inflow conditions (e.g. outlet-opening ratio), a sharp nozzle after end may also exhibit good stopping and astern operating performance. If the nozzle profile is more heavily rounded aft, ahead operating efficiency may be enhanced through a flow separation corner. Such flow separation corners may also be arranged on the forward ends to improve astern operating performance.

(c) Concentric form—oval inlet cross-section.   The theory of Amtsberg and the systematic experiments of van Manen investigated Kort nozzles in axial flow. This provides a good basis and reflected also practice up to the 1960s. Kort nozzles were pre-dominantly used in tugs which back then had very low C_B and predominantly axial propeller inflow. The situation in ocean-going ships today is different and the assumption of axial inflow is questionable. The side flanks of the nozzle may be opened and the nozzle axis oriented aft upwards to adjust for the different inflow direction.

A Kort nozzle thus adjusted for the inflow direction reduces power requirements considerably, but increases the costs of model testing and actually building the nozzle.

With simple-form nozzles, the opening is easily widened through the provision of a centro-symmetrical nozzle and subsequent installation of filling pieces. This 'Y-form' may also compensate an excessively small dihedral angle arising on height restriction grounds (Fig. 4.11).

For faired nozzles, an oval inlet can be designed at reasonable expense. 'Reasonable expense' means here that the nozzle is built in concentric form and then split, rather than two concentric nozzle parts and then assembling with intermediate pieces. The angle of the end of the inner part of the nozzle should be 2–3° towards the longitudinal axis. The propeller should always

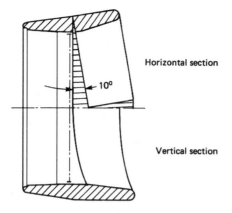

Horizontal section

Vertical section

Figure 4.11 Y-nozzle. Simple-form nozzle with lateral widening

have sufficient clearance (~2% of the propeller radius). The feasibility of installation of the combined nozzle must be checked.

(5) Nozzle axis direction

Nozzles are normally coaxially aligned with the propeller shaft. However, since the propeller incident flow is not quite coaxial, power requirement with the nozzle is frequently improved through matching of the nozzle axis to the inflow direction. For a twin-screw seagoing tug, for example, an aft-converging nozzle axis run with an angle of around 5° to the centreplane has proved particularly advantageous, despite aft divergence of the propeller shafts. For single-screw ships, an axis raked upwards going aft (Fig. 4.12), offers two advantages. Better adaptation to the flow is obtained, and, for a mounting penetrating the ship hull, better matching of the upper nozzle profile direction to the stern counter run can be obtained on the internal line of the nozzle. For cargo ships, optimum rake angles run from 5° to 7°. For nozzles with axes pointing aft upwards the design guidelines listed for Y nozzles apply.

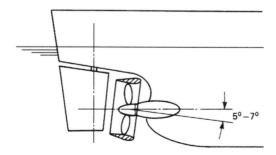

Figure 4.12 Single-screw ship with aft raked-up nozzle axis

(6) Profile length

Optimum nozzle profile length increases with thrust-loading coefficient. Nozzles are built with a length–internal diameter ratio of 0.4–0.8. The trend has been towards smaller lengths. At smaller lengths, a larger propeller diameter may be accomplished within a pre-determined vertical space. Profile length and cross-section shape are limited by strength and stiffness requirements. The profile length may be hydrodynamically optimized by Amtsberg's calculation procedure.

(7) Nozzle dihedral angle

Nozzle dihedral angle is the angle of the 'zero lift direction' or other profile reference line to the nozzle longitudinal axis. The dihedral angle may be optimized according to Amtsberg. At pre-selected nozzle total height, an increased dihedral angle means a restriction of propeller diameter or a more substantial distortion in the profile form in the lower part of the nozzle. Consideration must be given to this fact during selection of dihedral angle. Dihedral angle must also be considered in conjunction with profile form. If, to vary dihedral angle, the nozzle profile were only rotated, the outlet section would then be severely narrowed at large dihedral angles. At very small dihedral angles, there is the risk that the flow diffuser angle will become too large behind and the flow will become separated and eddying. Curved profiles, which avoid these difficulties, have so far been little studied and would also be too expensive to manufacture. The outlet angle to the longitudinal axis should be around 2° for Shushkin profiles and should not exceed 4° for faired profiles. If the dihedral angle is modified, the profile form must be matched to achieve a suitable outlet angle.

(8) Special devices for the deflection of objects flowing into the nozzle

On many cargo ships built without nozzles, such devices would have hydrodynamic and initial cost advantages. They are not used because operational disruptions are feared through jamming of the propeller in the nozzle, with particular apprehension about fouling by pieces of wood, ice, and stones drawn upwards from the bottom. Of the various ways to protect nozzles against inflowing objects, the preferred choice in practice is use of several annular grooves in the nozzle internal wall (Fig. 4.13). The boundary layer is thus

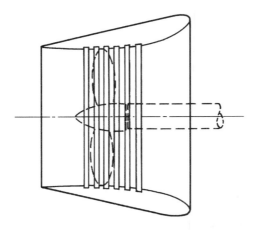

Figure 4.13 Nozzle with annular grooves in internal wall—longitudinal section at centre-line

thickened, with the result that inflowing objects are drawn inwards, leaving the gap between propeller blade tips and nozzle internal wall free.

(9) Cavitation and air entrainment

Since nozzles generate a strong depression field, cavitation and air entrainment can easily occur. Cavitation chiefly occurs at the nozzle internal wall in the proximity of the propeller. To avoid erosion damage, the internal wall is generally made of high-grade steel. Two measures are generally used to prevent air entrainment:

1. The nozzle is located as deep as possible. This requirement conflicts with the requirement for a larger diameter.
2. Arrangement of lateral skirts or a tunnel.

(10) Standard or Kaplan propeller

Kaplan propellers achieve better efficiencies in nozzles than propellers with elliptical contour lines. Kaplan propellers should not be run in steerable nozzles, since even greater gap widths are necessary. For ships operating in shallow waters, Kaplan propellers are more liable to be damaged by shingle than standard propellers. Therefore intermediate forms (Fig. 4.14) or standard propellers are used in these cases.

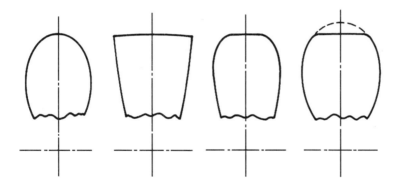

Figure 4.14 Blade tips of standard propeller, Kaplan propeller, and intermediate forms

Often errors are made in designing the Kort nozzle itself or its arrangement which can be easily avoided:

(a) Often the pressure side (exterior) of the nozzle is built as a cone which directly ends in a circle. The small curvature at the end is thus directly connected to an infinite radius of curvature of the straight section. The flow tends to separate due to this abrupt transition, at least at model scale. In full scale, flow separation is far less pronounced or absent. For model tests, it is thus advisable—or even necessary—to have a gradual change of curvature (Fig. 4.15). Comparative model tests show differences in efficiency of 6%.
(b) Accommodating the nozzle under the counter such that it penetrates the ship hull allows the maximum possible propeller radius and exploits the wake as far as possible. Furthermore, the attachment of the nozzle

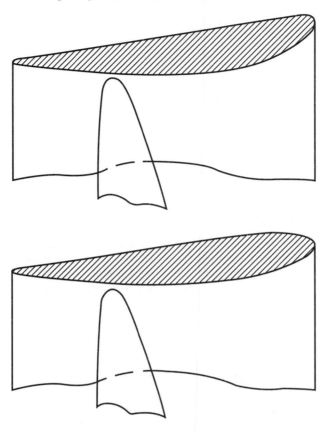

Figure 4.15 Strong change in curvature at nozzle entrance (top) and gradual change of curvature at nozzle entrance (bottom)

is very stable without using brackets which would increase resistance. The arrangement should ensure flow acceleration at the entrance in the upper region to avoid cavitation.

The nozzle contour declines downstream and the counter rises downstream. Therefore an intermediate section is necessary for strength reasons. This intermediate connection should not converge to a point, rather than a transom. Often, a hydrodynamically good solution is to fair the intermediate connection to the rudder contour (see also Fig. 4.26).

Saddle nozzles

The efficiency of a Kort nozzle can be described as a function of the thrust load coefficient. For full and slow ships, e.g. tankers, the thrust load coefficient may be locally in the upper quadrants more than 10 times as much as in the lower quadrants. This suggests locating the Kort nozzle only in the upper region where a high efficiency can be expected. Such a semi-nozzle is called a 'saddle nozzle' (Fig. 4.16), and has been successfully installed in models of

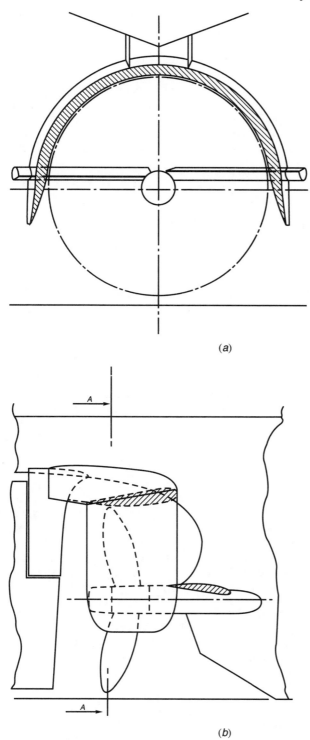

(a)

(b)

Figure 4.16 Saddle nozzle

cargo ships. Problems may occur with vibrations as the propeller tip enters the semi-circle. To reduce these vibrations, the radius of the semi-nozzle can be increased such that the propeller tip approaches the nozzle gradually. Another problem may be the reduced static strength and stiffness of the semi-nozzle. This may be improved by stiffening the entrance of the semi-nozzle with foils which may in addition give a pre-rotation to the propeller inflow.

The costs for saddle nozzles are higher than for complete Kort nozzles. Furthermore, classification societies require proofs of strength and vibrational characteristics. These proofs may be more expensive than the nozzle itself. Thus despite successful model tests, so far only one coastal freighter has been equipped with a saddle nozzle.

Further development

Kort nozzle have developed with the following objectives:

1. Better astern operating performance.
2. Simpler shaping.
3. Simpler manufacturing.
4. Greater safety against inflow or intake of shingle.
5. Efficiency enhancement by the 'Y-nozzle'.

4.6 Further devices to improve propulsion

Various devices to improve propulsion—often by obtaining a more favourable flow in the aftbody—have been developed and installed since the early 1970s, motivated largely by the oil crisis (Alte and Baur, 1986; Blaurock, 1990; Östergaard, 1996). Some of the systems date back much further, but the oil crisis gave the incentive to research them more systematically and to install them on a larger scale.

The Grim vane wheel

The Grim vane wheel consists of a relatively small propeller driven by the engine plant and a freely revolving propeller fitted on the downstream side, the inner part of which (behind the engine-driven propeller) acts as a turbine and the outer part as a propeller (Fig. 4.17) (Grim, 1966, 1980, 1982; Baur, 1985; Tanaka et al., 1990; Meyne and Nolte, 1991). This propulsion system has the following hydrodynamic advantages over normal single-propeller drive:

1. Substantial recovery of rotational energy.
2. Greater possible jet cross-section of vane wheel, since the low rpm rate and large number of blades enable smaller vertical clearances to be accepted.
3. Less resistance from rudder behind the vane wheel. This is reflected in the relative rotative efficiency.
4. Better stopping capability.

Moreover, the higher rpm rate associated with the smaller diameter of the engine-driven propeller improves the weight and cost of the propulsion unit. Grim proceeds from the assumption that the vane wheel is 20% larger in

Figure 4.17 Vane wheel system (figure from Bremer Vulkan)

diameter than the mechanically driven propeller. The system appears suitable for a wide range of conventional cargo ships, but only few actual installations have been reported.

Asymmetric aftbodies

Since 1982, several ships have been built with asymmetric aftbodies as patented by Nönnecke (1978,1987a,b) (Fig. 4.18). Model and full-scale tests indicate the following reasons for the power savings of 5–10%, especially for full hull forms (Collatz and Laudan, 1984; N. N., 1985; Nawrocki, 1989):

- Bilge vortex generation is reduced on the side with V-section characteristics (portside for clockwise turning propeller). Local separation is reduced on

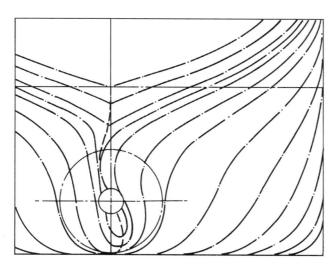

Figure 4.18 Hull sections of asymmetric aftbody

this side. This may lead to lower resistance for the asymmetric ship than the corresponding symmetrical ship in some cases.
• The pre-rotation induced by the hull improves the propeller efficiency. Rudder (and a vane wheel) reduce rotation as well.

Grothues spoilers

Cross-flows are often, but not always, observed in model tests investigating the ship flow near the propeller. This phenomenon decreases with distance from the hull. In addition, bilge vortices appear (Fig. 4.19). The cross-flow usually has a thickness comparable to that of the boundary layer. Cross-flows appear predominantly in ships with stern bulb, high B/T, high C_B and low speed. Cross-flows disturb the propeller inflow and reduce the propeller efficiency.

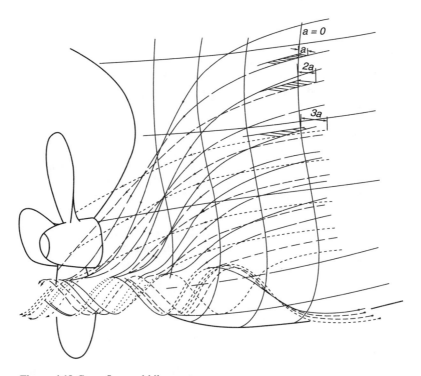

Figure 4.19 Cross-flow and bilge vortex

Grothues-Spork (1988) proposed spoilers—fitted before the propeller on both sides of the stern post—to straighten horizontally the boundary layer flow right before the propeller, thus creating direct thrust and improving the propeller efficiency. He used parts of a cylindrical surface such that they divert more strongly near the hull and less so further out. These fins are called Grothues spoilers (Fig. 4.20).

Power savings measured in model tests were:

Tankers and bulkers, fully loaded	up to 6%
Tankers and bulkers, in ballast	up to 9%

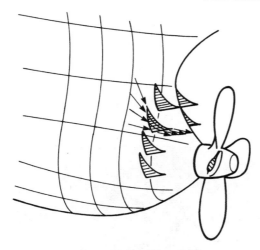

Figure 4.20 Grothues spoilers in principle

Ships of medium fullness with $B/T < 2.8$ up to 6%

Fine vessels with small B/T up to 3%

Special investigations on the spatial flow conditions in the propeller post region have to be made for the determination of shape, position and number of spoilers. The expense of manufacturing and fitting spoilers is generally low.

The wake equalizing duct

In the following, we will first treat wake equalizing ducts for single-screw ships. The wake equalizing duct (WED) is a ring-shaped flow vane with foil-type cross-section fitted to the hull in front of the upper propeller area (Fig. 4.21) (Schneekluth, 1985, 1989; Stein, 1983, 1996; N. N., 1986, 1992; Renner, 1992; Steirmann, 1986; Xian, 1989). The WED is by far the most frequently installed propulsion improving device (Meyne, 1991) (Table 4.1). In contrast to the Kort nozzle, which shrouds the propeller, these ducts are less than half as big in diameter and section length and are arranged in the wake. They are fitted to the hull in the form of two half-ring ducts in front of the propeller. Their upper ends may be integrated to the hull ahead of the stern frame or they may extend into the stern aperture, in which case the gap at the trailing edge aft of the stern frame is given a horizontal filling. WEDs consist usually of two centro-symmetric halves which are connected by straight foil-type parts to the hull. For an asymmetric stern fitting a half-ring duct on only one side can be more beneficial than the double-sided arrangement. The duct

Table 4.1 Installations of propulsion improving devices up to 1991

Wake equalizing duct	>500
Asymmetric aftbody	75
Vane wheel	60
Grothues spoilers	35

136

Figure 4.21 Wake equalizing duct (WED)

is most effective on the side with larger curvature of the waterlines. The basic principle underlying the application of this device is that the flow creates a circulation around the foil section of half-ring ducts which accelerates the flow in the area enclosed by them and retards it in their outer environment. Thus, such a nozzle channels the flow in the upper quadrants where it matters most. The inward-directed circulation guides the water into the duct, and ahead of it presses the flow on to the hull. The flow is then better attached to the hull and separation prior to the duct is reduced (Fig. 4.22).

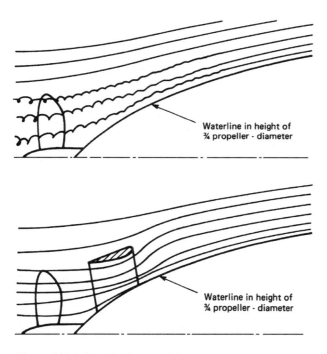

Figure 4.22 Schematic diagram of flows
Top: flow along a waterline at a height of about 3/4 propeller diameter. In stern region separation occurs
Below: flow with duct, no separation

The WED is characterized by the following parameters:

- Inner diameter (43–44% of propeller diameter).
- Chord length (50–70% of inner diameter).
- Profile section shape (special, not corresponding to any standards).
- Angle of outline cone.
- Angle of axis of half rings against the longitudinal and transverse planes of the ship, which have different settings for port and starboard sides.
- Distance of axes from each other—taken at the exit plane.
- Distance of WED from propeller.

A normal longitudinal section across the duct explains the circulation effect relating to the speed distribution in the upper and lower halves of the propeller.

The inflow of the propeller is accelerated in the upper region where it is slow, corresponding to the fuller form of the ship, and in the lower region, where the speed of inflow is normally higher, it will be retarded. In practice the average and effective wake will hardly be changed (Fig. 4.23). In accelerating nozzles and ducts the open cross-section at the trailing edge is usually smaller than that at the leading edge. This often may not be so in WEDs. The flow in the WED region has divergent flowlines due to the ship hull form. The WED decreases this divergence by locally accelerating the flow in this region.

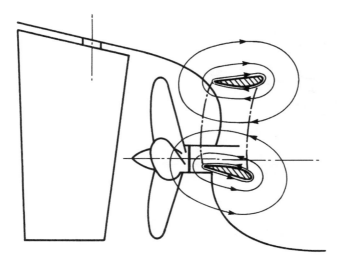

Figure 4.23 Circulation in vertical direction

Advantages of application

The main advantage lies in power savings resulting from various effects:

1. Improved propeller efficiency from more axial flow and more uniform velocity distribution over the disc area. The former effect dominates. Measurements on a containership model show that the angle of inward inclination of flow in the plane behind the duct is reduced from as much as 20° to about 7° to the longitudinal axis of the ship. The asymmetrical arrangement of half ducts gives a rotational direction to the water entering the propeller, which is opposite to that which the propeller will impart. Thus the loss from rotation energy in the propeller wake is less.
2. Reduction of flow separation at the aftbody. This effect is strong and reduces resistance and the thrust deduction fraction.
3. Lift generation with a forward force component on the foil section, similar to but weaker than that in the Kort nozzle (Fig. 4.24).
4. The nozzle axes are oriented such that the propeller inflow is given a slight pre-rotation which counteracts the propeller rotation.
5. Improved steering qualities from more straightened flow to the rudder. In spade rudders the longer upper sections become more effective because of the higher flow velocity.

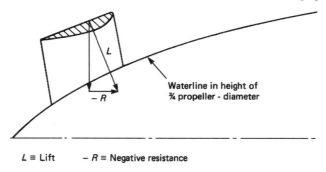

$L \equiv$ Lift $- R \equiv$ Negative resistance

Figure 4.24 Schematic diagram of lift with forward force component on duct

6. Improved course-keeping ability from increased lateral plan area aft.
7. No constructional changes and no modifications in propeller design are involved when the duct is fitted to an existing ship.
8. Possibility to integrate devices for ice protection to propeller. Even without special ice protection, ducts protect propellers. Up to 1997, almost 900 ducts had been installed, many in ships on ice-infested routes. No damage to ducts has been reported and ice-damage to propellers has been reduced.
9. Reduction of propeller-excited vibrations from decreased propeller tip loading in upper quadrants to less than half the amplitudes. This allows reduction of propeller clearances in new designs. Reduced vibrations have in practice also decreased malfunctions of electronic equipment. The reduction in vibration amplitudes by the WED is easily explained by the velocity distribution. Larger inflow velocity means smaller angle of attack α between profile zero lift position and inflow direction (Fig. 4.25). The hydrodynamic forces and thus pressure impulses are roughly proportional to the angle of attack for small angles of attack. The WED also smooths the torque and thus reduces the tendency for torsional vibrations.

The power savings can be used to obtain higher speed. For a given speed, the power savings are converted to a lower rate of rotation.

The WED leads to a differently distributed inflow to the propeller, but not a higher average inflow velocity. In fact, the additional friction in the duct increases the wake fraction by some 0.01. This hardly changes the optimal propeller pitch. Thus, an often desired correction in propeller pitch cannot be achieved by a WED.

The positive effect of the WED in power saving is most evident in the speed range up to 23 knots. Generally the power gain increases with speed. The relationship of power gain to speed shows an analogous behaviour to that of effective wake and speed. The maximum advantage is obtained mostly at the full-load condition. At ballast draft the gain is smaller, mainly because of the stern trim associated with this draft condition. Model tests have determined a maximum energy saving of 14% at the same speed in several cases. In all cases (i.e. with or without duct) the results were converted by the standard procedure without any corrections for scale effects of the duct. The WED also offers the possibility of injecting air at the propeller for the purpose of reducing

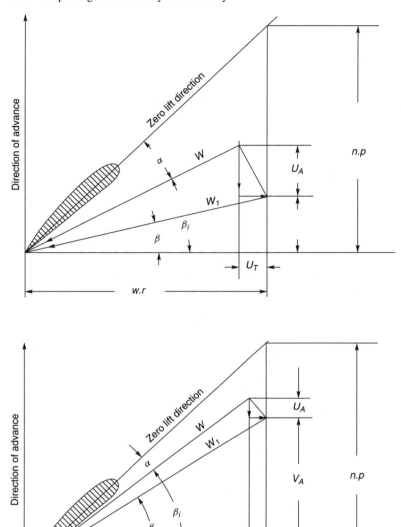

Figure 4.25 Velocities and angle of attack at propeller for low (top) and high (bottom) inflow velocity.

v_A axial propeller inflow velocity
ω_r radial velocity
u_A additional velocity in axial direction
u_T additional velocity in radial direction
w resultant inflow velocity
n propeller rpm
P pitch
α profile angle of attack

the impulse impingement on the hull surface and of reducing cavitation. This option has never been used in practice, although it is simpler than injecting air via a canal system in the propeller blades.

Cavitation

Unlike ducted propellers, which are hardly used in ocean vessels due, in part, to problems of tip clearances and cavitation, the WED does not pose such problems. The more uniform flow into the propeller reduces the dangers of propeller cavitation. The duct itself is less exposed to this problem than the rudders because of the considerably lower flow velocity in the wake at its location, which is often less than half the ship speed. Another advantage here is that normally the WED is considered for moderate speed vessels with block coefficient above 0.6; fast ships, which tend to have cavitation problems, are less suited to its use.

Scope of application

To reduce possible propeller-excited vibrations and to improve hull efficiency, modern designs often incorporate stern bulbs, bigger propeller tip clearances and slender run of waterlines in the region of the upper quadrants. For concave waterlines in the region of the 'critical waterline', i.e. half a propeller radius above the shaft height, the onset of flow separation may be too far ahead to be captured by the nozzle circulation. The nozzle cannot reverse separation once it has started. If in this case the nozzle is placed further ahead than usual, the interaction with the propeller deteriorates. The effect of not capturing the flow separation is mainly a problem for model tests, as flow separation is shifted further aft in full scale.

The bases for evaluation of economic gains are expected power savings from comparative model tests or from experience gained from other vessels fitted with the nozzle. Data required for a preliminary assessment consist of hull lines fullness and details of the propeller and its configuration.

In newbuilds, it is recommended that model tests be extended to include duct variants to determine the best arrangement and attainable gains, because these tests involve relatively low additional costs. For fitting to an existing ship, where a model has to be manufactured specially for this purpose, model testing can be rather costly.

Cost aspects

In newbuilds the costs of the duct can be lower than those costs saved by choice of a smaller engine, made possible by the power savings. Even when a suitable, next smaller engine is not available the shipowner still saves fuel, although the initial investment is then slightly higher. The investment for fabrication and fitting is invariably recovered in 6–20 months, depending on ship form and fuel price (Stiermann, 1986).

Integration in ship design

The interaction between the ship and the duct raises the question of whether there is further scope for improvement by adopting the aftbody design for duct integration. In ships with $C_B > 0.6$, flow separation in the stern area cannot be completely avoided. When duct integration is envisaged, it is better to locate these areas in the duct region, where it effectively reduces flow separation, i.e. the waterlines ahead of the duct should not be kept hollow but should have their greatest slope here. The increase in thrust deduction fraction from the greater waterline slope is more than compensated by the increased effectiveness of the duct. Similarly, the horizontal propeller blade clearance from the stern frame need not be kept wide to avoid undesirable effects from propeller action. Adequate smaller clearances, such that the duct does not completely extend into the aperture, also improve the duct effectiveness in respect of separation.

For new designs, the WED offers additional advantages.:

+ The ship hull can be kept simpler. The stern bulb can be built less pronounced and the counter can be placed lower. Concave waterlines at the height of the WED are not necessary, thus the hull is cheaper to produce and the resistance lower.
+ Simpler propellers with fewer blades and less skew. The propellers can be more highly loaded at the tips. Thus the propellers are cheaper, yet more efficient.

Conversion of results from model tests

Unfortunately, even computations based on 'Navier–Stokes' codes (see Section 2.11), have not yet been able to determine the power savings from WEDs. Accurate prediction of flow separation remains a problem. One still reverts to model testing or sea trials. If no model tests are envisaged prior to the installation of the duct, comparative test data to cover most cases can be used. Estimates based on comparable ships are generally in respect of design draught and speed. On the other hand it is commonly not possible to predict, without model tests, the amount by which the power savings will change with variation of speed, draught or trim. In a ship model with WED, significant scale effects occur, about which quantitatively little is known. These are in favour of the full-sized ship so that actual gains for the ship may be 2–3% higher than those predicted in model tests. This difference is not explained by the higher frictional resistance of the duct, as this would contribute only 0.3–0.5% to the total power prognosis. Sea trials and data obtained from long-term operation confirm power savings up to 8% on average over the whole range of service conditions in respect of draft and speed. For conversion of model test results to full scale, three factors act in favour of the full-sized ship, but are generally not taken into consideration for predictions given in the test reports.

1. Scale effects

The difference in frictional resistance coefficient for the duct in model and full scale is considerably higher than that allowed for by frictional deduction allowance for skin friction of ship and model. The difference in friction resistance coefficient cannot be ascertained easily because the flow velocity around the duct is not known unless measured. Another scale effect is due to lack of similarity in boundary layer thickness. Due to the relatively thinner boundary layer on the ship, the volume of water passing through the duct is bigger. As a third scale effect, the component of resistance from flow separation can be

different in model and in ship. The separation effect is slightly more exaggerated in the model, implying that the possible reduction in separation can be
greater here. The difference in flows at model and full scale is schematically
displayed in Fig. 4.26. This separation effect is, unusually, in favour of the
model.

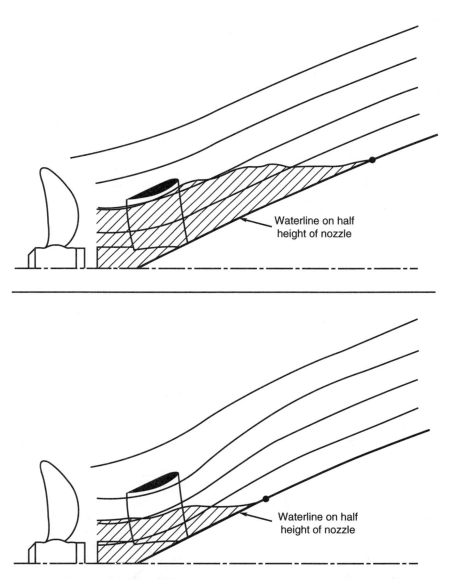

Figure 4.26 Principle of different WED effect in full scale (top) and model scale (bottom);
flowlines and areas of separation

2. Model similarity

In model tests the ducts are fitted to the ship model on shafts so that the setting
of vertical and horizontal axis angles can be varied to determine the optimal

arrangement. The additional resistance of the shafts and the gaps at the connection of half-ring ducts to the ship model can increase the resistance, thus reducing the effectiveness of the ducts in the model.

3. Seastate influence

Comparisons between the ship with and without WED refer to smooth water performance. Model tests with a containership in smooth water and in regular waves show an additional power saving in seastates, amounting to about 3–4% for the model with duct, as against the model without it. The wavelength in these tests was from 0.5 to 1.5 ship length and the wave height was 3% of ship length.

Construction, fitting and mass

Construction and pre-fabrication of half-ring ducts is similar to that for the Kort nozzle. For practical reasons, the plate thickness in fabrication is much greater than strength considerations demand. Connection to the stern frame structure usually requires no additional internal stiffening of the stern frame. All WEDs so far have been built using welded construction. Shell plate thickness ranges from 7–14 mm for ducts of 1–3 m diameter. Fitting the WED to the ship in dock takes only a few days. The Thyssen Nordsee shipyards in Emden have developed a method to fit WEDs on floating, trimmed ships from a pontoon.

The weight of WED, m_D [t] with a profile length of half the inner diameter D_i [m] can be approximated by

$$m_D \approx D_i^{2.3} \cdot 0.48 - \frac{0.1}{D_i}$$

For a profile length of $0.65 D_i$ we have:

$$m_D \approx D_i^{2.5} \cdot 0.47 + 0.1$$

The equations are for both half rings together and for $D_i > 1.2$ m.

WED for twin-screw ships

WEDs have also been installed successfully in twin-screw ships. Twin-screw ships usually feature more uniform wakes than single-screw ships. The wake affects the twin-screw propeller predominantly on the side near the hull and in the flow region behind the shaft brackets. The WED can equalize the wake also in this case, but should be concentric around the shaft, accelerating the flow in an arc from approximately 90° to 130° in the region of strong wake. In the region between WED and hull the flow will be slowed down. Power savings are thus derived from an increased propeller efficiency due to equalized wake and a reduced hull friction resistance behind the WED. Winglets at both tips of the WED segment may yield further power savings. The main effect of WED for single-screw ships, the reduction of separation, is not applicable for twin-screw ships. Yet installations in passenger ships showed speed improvements around half a knot. We cannot yet give a physical explanation for this effect. For new designs of twin-screw ships, WEDs can reduce the resistance of appendages, as shaft brackets can be kept shorter and more slender. Also, the water-immersed part of the shaft can be kept shorter or run at a lower angle towards the flow. If the propeller is not arranged closer to the hull, the propeller diameter may be increased. Figure 4.27 shows qualitatively the flow field for a twin-screw ship with and without a WED. WED for twin-screw ships are usually welded to the shaft brackets without further connecting elements.

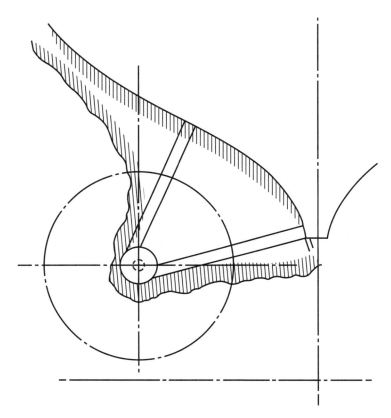

Figure 4.27 Effect of WED on flow in twin-screw ships

Combination of devices

Devices to improve propulsion have also been successfully combined. However, savings given for individual systems will not add up completely for combinations of systems. The estimates of total efficiencies which can be obtained given below are just guidelines. Also, in practice such combinations are rarely found as the high complexity of the systems introduces additional initial and sometimes operating (maintenance and repair) costs. Designers therefore generally favour—at least in times of relatively low fuel costs—simple solutions involving at most one system to improve propulsion, e.g. a WED.

Grim vane wheel and asymmetric aftbody

Combinations have been installed (e.g. Kringel and Nolte, 1985; Spruth *et al.*, 1985). As both systems are based on the recovery of rotational energy, the

combination will give only 65–75% of the sum of the savings expected for
each of the systems.

Grim vane wheel and Grothues spoilers

This combination is possible and has been tested on different ship types. The
total efficiency improvement is 75–85% of the sum of the individual savings,
as the resistance decrease given by the spoilers reduces the efficiency of the
vane wheel.

Grim vane wheel and WED

The situation is similar to that for the combination vane wheel/spoiler systems,
but the WED gives a slight additional rotation in the flow, reducing total
savings to 70–80% of the sum of individual savings.

Grothues spoilers and asymmetric aftbody

Model tests for this combination were not encouraging as both systems aim
to reduce bilge vortex formation.

Grothues spoilers and WED

Cross-flows, which motivated the development of spoilers, also decrease the
WED efficiency. For ships featuring cross-flows, Grothues spoilers in front of
the WED increase efficiency and decrease propeller vibrations. In these cases,

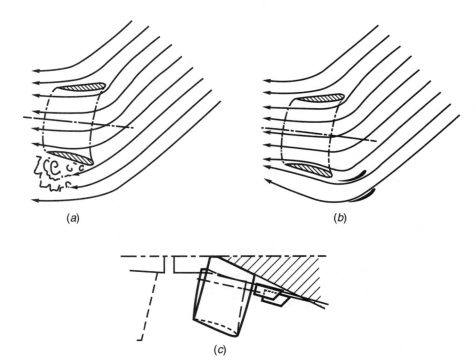

(a)   (b)

(c)

Figure 4.28 Cross-flow near hull without (left) and with (right) Grothues spoilers in front of
WED

two spoilers in front or slightly below the WED–hull intersection are usual. Figure 4.28 shows, in principle, the effect of the spoilers. Spoilers used in combination with WED have relatively thick profiles and large hull intersections. As with the WED, they do not require any further stiffeners. More than 180 combinations of WED and spoilers had been reported by 1997.

WED and asymmetric aftbody

This combination has been realized several times (Schneekluth, 1985). In this combination, the duct is placed on one side of the ship, namely the 'upper concave' side, i.e. the starboard side for a clockwise turning propeller. Quantification of the expected total improvement of efficiency is difficult: this will have to be determined individually by model tests.

Grim vane wheel, asymmetric aftbody and WED

The combination is possible (Kringel and Nolte 1985; Spruth *et al.*, 1985), however, the high complexity of these combined systems has prevented widespread application.

4.7 References

ALTE, R. and BAUR, M. v. (1986). Propulsion. *Handbuch der Werften*, Vol. **XVIII**, Hansa, p. 132

AMTSBERG, H. (1950). Entwurf von Schiffdüsensystemen (Kortdüsen)—Praktisches Auswahlverfahren für optimale Düsensysteme. *Jahrbuch Schiffbautechn. Gesellschaft*, p. 170

BAUR, M. v. (1985). Grim'sches Leitrad seit zwei Jahren an Handelsschiffen im Einsatz. *Hansa*, p. 1279

BLAUROCK, J. (1990). An appraisal of unconventional aftbody configurations and propulsion devices. *Marine Technology* **27/6**, p. 325

BUSSEMAKER, O. (1969). Schottel-Antriebe. *Hansa*, p. 149

COLLATZ, G. and LAUDAN, J. (1984). Das asymmetrische Hinterschiff. *Jahrbuch Schiffbautechn. Gesellschaft*, p. 149

GRIM, O. (1966). Propeller und Leitrad. *Jahrbuch Schiffbautechn. Gesellschaft*, p. 211

GRIM, O. (1980). Propeller and vane wheel. *Journal of Ship Research* **24/4**, p. 203

GRIM, O. (1982). Propeller und Leitrad auf dem Forschungsschiff 'Gauss'. Ergebnisse und Erfahrungen. *Jahrbuch Schiffbautechn. Gesellschaft*, p. 411

GROTHUES-SPORK, H. (1988). Bilge vortex control devices and their benefits for propulsion. *International Shipbuilding Progress* **35**, p. 183

HENSCHKE, W. (1965). *Schiffbautechnisches Handbuch* Vol. **1**. Verlag Technik, Berlin, p. 562

HORN, F. (1940). Beitrag zur Theorie ummantelter Schiffsschrauben. *Jahrbuch Schiffbautechn. Gesellschaft*, p. 106

HORN, F. (1950). Entwurf von Schiffsdüsensystemen (Kortdüsen)—Theoretische Grundlagen und grundsätzlicher Aufbau des Entwurfsverfahrens. *Jahrbuch Schiffbautechn. Gesellschaft*, p. 141

ISAY, W. H. (1964). *Propellertheorie—Hydrodynamische Probleme*. Springer

KRINGEL, H. and NOLTE, A. (1985). Asymmetrisches Hinterschiff, Zuströmdüse und Leitrad auf den Container-Mehrzweck-Frachtschiffen Arkona und Merkur Island. *Hansa*, p. 2472

LINDGREN, H., JOHNSON, C. A. and DYNE, G. (1968). Studies of the application of ducted and contra-rotating propellers on merchant ships. *7th Symposium of Naval Hydrodynamics*, Office of Naval Research, p. 1265

MEYNE, K. J. (1991). 500 Schneekluth-Düsen innerhalb von fünf Jahren installiert. *Hansa*, p. 832

MEYNE, K. and NOLTE, A. (1991). The Grim wheel. Cavitation and tip vortex. Observations and conclusions. *Schiffstechnik*, p. 191

MUNK, T. and PROHASKA, C. W. (1968). *Unusual Propeller Arrangements*. Hy A Lungby, Denmark

NAWROCKI, S. (1989). The effect of asymmetric stern on propulsion efficiency from model test of a bulk carrier. *Schiff + Hafen* **10**, p. 41

N. N. (1985). Development of the asymmetric stern and service results. *Naval Architect*, p. E181

N. N. (1986). Wake-equalizing ducts for twin-screw ships. *Naval Architect*, p. E147

N. N. (1992). Schneekluth wake-equalising ducts for twin-screw ships. *Naval Architect*, p. E473

N. N. (1993). Two VLCCs with contra-rotating propellers in service. *Naval Architect*, p. E444

NÖNNECKE, E. A. (1978). Reduzierung des Treibstoffverbrauches und Senkung der Betriebskosten der Seeschiffe durch propulsionsverbessernde Maßnahmen. *Hansa*, p. 176

NÖNNECKE, E. A. (1987a). Schiffe mit treibstoffsparendem asymmetrischen Heck—Anwendungen und Erfahrungen. *Schiff + Hafen* **9**, p. 30

NÖNNECKE, E. A. (1987b). The asymmetric stern and its development since 1982. *Shipbuilding Technology International*, p. 31

ÖSTERGAARD, C. (1996). Schiffspropulsion. *Technikgeschichte des industriellen Schiffbaus in Deutschland* **Vol. 2**, Ed. L. U. Scholl. Ernst Kabel Verlag, p. 65

PAETOW, K. H., GALLIN, C., BEEK, T. v. and DIERICH, H. (1995). Schiffsantriebe mit gegenläufigen Propellern und unabhängigen Energiequellen. *Jahrbuch Schiffbautechn. Gesellschaft*, p. 451

PHILIPP, O., HEINKE, H. J. and MÜLLER, E. (1993). Die Düsenform—ein relevanter Parameter der Effizienz von Düsen-Propeller-Systemen. *Jahrbuch Schiffbautechn. Gesellschaft*, p. 242

PIEN, P. C. and STROM-TEJSEN, K. (1967). A Proposed New Stern Arrangement. Report 2410. Naval Ship Research and Development Center (NSRDC), Washington, D.C.

RENNER, V. (1992). Schneekluth-Düsen gibt es jetzt auch für Doppelschrauber mit Wellenbockarmen. *Schiff + Hafen* **10**, p. 166

SAVIKURKI, J. (1988). Contra-rotating propellers. *Hansa*, p. 657

SCHNEEKLUTH, H. (1985). Die Zustromdüse—alte und neue Aspekte. *Hansa*, p. 2189

SCHNEEKLUTH, H. (1989). The wake equalizing duct. *Yearbook of The Institute of Marine Engineers*

SPRUTH, D., WOLF, H., STERRENBERG, F. *et al.* (1985). BV 1000—ein neues Typschiff der Bremer Vulkan AG. Neubauten für wirtschaftlicheren Containertransport. *Schiff + Hafen* **8**, p. 23

STEIN, N. von der (1983). Die Zustromausgleichsdüse. *Hansa*, p. 1953

STEIN, N. von der (1996). 12 Jahre Schneekluth-Zustromdüse. *Hansa*, p. 23

STIERMANN, E. J. (1986). Energy saving devices. Marin-Report **26**, Wageningen

TANAKA, M., FUJINO, R. and IMASHIMIZU, Y. (1990). Improved Grim vane wheel system applied to a new generation VLCC. *Schiff + Hafen* **10**, p. 146

VAN MANEN, J. D. and SENTIC, A. (1956). Contra-rotating propellers. *International Shipbuilding Progress*, **3**, p. 459

XIAN, P. (1989). Strömungsmechanische Untersuchungen der Zustromdüse. Ph.D. thesis, TU Aachen

5

Computation of weights and centres of mass

All prediction methods should be calibrated using data from comparable ships. This allows the selection of appropriate procedures for a certain ship type (and shipyard) and improves accuracy.

The prediction of weights and centres of mass is an essential part of ship design. A first, reasonably accurate estimate is necessary for quoting prices. A global price calculation is only acceptable for follow-up ships in a series, otherwise the costs are itemized according to a list of weight groups. In many cases, it is still customary to calculate not only the material costs, but also the labour costs based on the weight of the material.

The largest single item of the ship's weight is the steel weight. Here, first the installed steel weight (net weight) is estimated. Then 10–20% are added to account for scrap produced, for example, in cutting parts. Modern shipyard with accurate production technologies and sophisticated nesting procedures may use lower margins.

The displacement Δ of the ship is decomposed as

$$\Delta = \Delta_L + W_{dw} = (W_{StR} + W_{StAD} + W_o + W_M + W_R) + W_{dw}$$

The symbols denote:

Δ_L weight of ship without payload (light ship)
W_{StR} weight of steel hull
W_{StAD} weight of steel superstructure and deckhouses
W_o weight of equipment and outfit
W_M weight of engine (propulsion plant)
W_R weight margin
W_{dw} total deadweight including payload, ballast water, provisions, fuel, lubricants, water, persons and personal affects

The exact definitions of the individual weight contributions will be discussed in subsequent sections. All weights will be given in [t], all lengths in [m], areas in [m^2], volumes in [m^3].

For cargo ships, the displacement may be globally estimated using the ratio $C = W_{dw}/\Delta$ and the specified deadweight W_{dw}. C depends on ship type, Froude number and ship size. This procedure is less appropriate for ships where the size is determined by deck area, cargo hold volume or engine power, e.g. ferries, passenger ships, tugs and icebreakers.

For cargo ships $C \approx 0.66$

For tankers $C \approx 0.78 + 0.05 \cdot \max(1.5, W_{dw}/100\,000)$

The height of the centre of mass can be similarly estimated in relation to the depth D or a modified depth D_A:

$$\overline{KG} = C_{KG} \cdot D_A = C_{KG} \cdot \left[D + \frac{\nabla_A + \nabla_{DH}}{L_{pp} \cdot B} \right]$$

∇_A is the superstructure volume and ∇_{DH} the volume of the deckhouses. D_A is depth corrected to include the superstructure, i.e. the normal depth D increased by an amount equal to the superstructure volume divided by the deck area. Values in the literature give the following margins for C_{KG}:

passenger ships   0.67–0.72

large cargo ships   0.58–0.64

small cargo ships   0.60–0.80

bulk carrier   0.55–0.58

tankers   0.52–0.54

Table 5.1a Percentage of various weight groups relative to light ship weight

		dw/Δ [%]	W_{St}/Δ_L [%]	W_o/Δ_L [%]	W_M/Δ_L [%]
cargo ship	5000–15 000 tdw	60–80	55–64	19–33	11–22
coastal cargo ship	499–999 GT	70–75	57–62	30–33	9–12
bulker	20 000–50 000 tdw	74–80	68–79	10–17	12–16
bulker	50 000–150 000 tdw	80–87	78–85	6–13	8–14
tanker	25 000–120 000 tdw	65–83	73–83	5–12	11–16
	≥200 000 tdw	83–88	75–83	9–13	9–16
containership	10 000–15 000 tdw	60–76	58–71	15–20	9–22
	20 000–50 000 tdw	60–70	62–72	14–20	15–18
ro-ro ship	≤16 000 tdw	50–60	65–78	12–19	10–20
reefer	300 000–600 000 cu ft	45–55	51–62	21–28	15–26
ferry		16–33	56–66	23–28	11–18
trawler	44–82 m	30–58	42–46	36–40	15–20
tug	500–3000 kW	20–40	42–56	17–21	38–43

Table 5.1b Height of centres of mass above keel [% height of top-side deck above keel]

		for W_{St}	for W_o	for W_M	light ship
cargo ship	≥5000 tdw	60–68	110–120	45–60	70–80
coastal cargo ship	≥499 GT	65–75	120–140	60–70	75–87
bulker	≥20 000 tdw	50–55	94–105	50–60	55–68
tanker	≥25 000 tdw	60–65	80–120	45–55	60–65
containership	≥10 000 tdw	55–63	86–105	29–53	60–70
ro-ro ship	≥80 m	57–62	80–107	33–38	60–65
reefer	≥300 000 cu ft	58–65	85–92	45–55	62–74
ferry		65–75	80–100	45–50	68–72
trawler	≥44 m	60–65	80–100	45–55	65–75
tug	≥500 kW	70–80	100–140	60–70	70–90

| trawlers | 0.66–0.75 |
| tugs | 0.65–0.75 |

Table 5.1 compiles the percentage of various weight groups and the centres of mass.

5.1 Steel weight

The 'steel weight' is regarded as the quantity of rolled material processed in the actual manufacture of the ship. This includes plates, sections, castings for the stern and tail-shaft brackets and the processed weld metal. More exact demarcations *vis-à-vis* other weight groups differ between shipyards. In particular, there are the following components, classed partly under 'steel' and partly under 'equipment and outfit':

1. Steel hatchway covers.
2. Masts.
3. Rudder shell.
4. Container guides.

Procedures for calculating steel weight

By far the greatest part of the hull weight is made up by the steel weight. For this reason, more precise weight calculation methods are applied to better determine this quantity, even though the weight group 'equipment and outfit' may only be approximated.

The procedures to calculate steel weight are based on the steel weights of existing ships or on computed steel weights obtained from construction drawings produced specially for the procedure. Both cases require interpolation and extrapolation between the initial values of the parameters. The procedures ascertain, either:

1. The overall quantity of steel.
2. Only the hull steel or the steel used in the superstructure and deckhouses.
3. Individual larger weight groups—e.g. outer shell, decks, double bottom—from which the total steel weight can be formed.

The main input values are the main dimensions, number of decks and construction type.

Empirical methods developed for conventional ships cannot be applied to unconventional ships. Then the following procedure—the original approach is credited to Strohbusch and dates back to 1928—is recommended:

1. Calculation of hull steel weight per cross-sectional area or rate per metre ship length for some prominent cross-sections (Fig. 5.1).
2. Plotting of 'weight per unit length' over the ship's length.
3. Determination of the area below the weight curve.
4. Addition to the weight thus determined of individual weights not included in the running weight per unit length.

The area below the curve in Fig. 5.2 represents the weight.

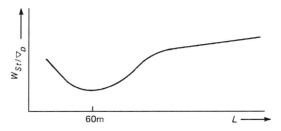

Figure 5.1 Specific steel weight in relation to length

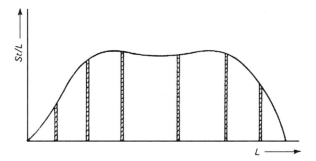

Figure 5.2 Distribution of hull steel weight over the ship's length (for a ship with sheer)

Coefficient methods

Steel weight calculation procedures are often based on formulae of the form:

$$W_{St} = L^a \cdot B^b \cdot D^c \cdot C_B^d \cdot e$$

where a–e are constants. Some procedures omit the C_B term. Then the result relates to 'type-conventional' C_B values. Some procedures are only implicitly expressed in terms of the main dimensions.

Although most methods do not give details of construction, e.g. number of decks, they can nevertheless be sufficiently precise, when relating to a specific ship type and to a particular size range, and are still used in practice at least for a first estimate in the design spiral (Hollenbach, 1994). Moreover, it is assumed that the normal main dimension relationships are maintained, since the exponent of the length changes with variation in length. Carstens (1967) presents a more sophisticated approach also including such details as the number of decks.

Generally, coefficient methods should be calibrated using modern comparable ships. For better accuracy, differences in details of the steel structure and dimensioning loads for project ship and comparison ship should be taken into account. Some examples demonstrate the importance of this point:

Differences in structural design of tanker bulkheads:

Tanker with corrugated bulkheads, spec. cargo
    weight 1.85 t/m³   4420 t steel

Tanker with welded stiffeners   4150 t steel

Differences in dimensioning loads for tanker bulkheads:

Tanker with 10 tanks, spec. cargo weight 1.10 t/m^3 3880 t steel

Tanker with 10 tanks, spec. cargo weight 1.55 t/m^3 4020 t steel

Tanker with 24 tanks, spec. cargo weight 2.10 t/m^3 4740 t steel

Differences in ice strengthening for tanker with 10 tanks, spec. cargo weight 1.10 t/m^3:

Tanker, strengthened for GL E3, no intermediate sections   440 t

Tanker, strengthened for GL E3, intermediate sections   220 t

Tanker, strengthened for GL E3, intermediate sections, HT steel   175 t

Differences in structural design and loads on ro-ro decks:

Ro-ro ship, mild steel, 55 t axle load, no supports   5700 t

Ro-ro ship, mild steel, 55 t axle load, 2 rows of support   4970 t

Ro-ro ship, HT steel, 17 t axle load, no supports   4100 t

Computer-aided design methods allow determination of the areas of plates on the hull and bulkheads quickly and accurately. Also specific weights (per area) of stiffened plates can be quickly determined using the dimensioning tools of classification societies which consider the distance between stiffeners, loads and material.

Some special methods

Containerships

Miller (1968):

$$W_{St} = 0.000435(L \cdot B \cdot D)^{0.9} \cdot (0.675 + C_B/2)$$
$$\cdot [0.00585((L/D) - 8.3)^{1.8} + 0.939]$$

Dry cargo vessels

Kerlen (1985):

$$W_{St} = 0.0832 \cdot X \cdot e^{-5.73X \cdot 10^{-7}} \text{ with } X = \frac{1}{12}L_{pp}^2 \cdot B \cdot \sqrt[3]{C_B}$$

Watson and Gilfillan (1977):

$$W_{St} = C_B^{2/3} \cdot \frac{1}{6}L \cdot B \cdot D^{0.72} \cdot [0.002(L/D)^2 + 1]$$

Tankers

Det Norske Veritas (1972):

$$W_{St} = \Delta[\alpha_L + \alpha_T(1.009 - 0.004 \cdot (L/B)) \cdot 0.06 \cdot (28.7 - (L/D))]$$

where:

$$\alpha_L = [(0.054 + 0.004\,L/B) \cdot 0.97]/[0.189 \cdot (100\,L/D)^{0.78}]$$
$$\alpha_T = 0.029 + 0.00235 \cdot \Delta/100\,000 \qquad \Delta < 600\,000\,\mathrm{t}$$
$$\alpha_T = 0.0252 \cdot (\Delta/100\,000)^{0.3} \qquad \Delta > 600\,000\,\mathrm{t}$$

Range of validity:

$$L/D = 10\text{--}14, \quad L/B = 5\text{--}7, \quad L = 150\text{--}480\,\mathrm{m}$$

Normal steel; superstructure and deckhouses are not included.
   Sato (1967):

$$W_{St} = (C_B/0.8)^{1/3} \cdot [5.11 \cdot L^{3.3} \cdot B/D + 2.56 \cdot L^2(B + D)^2]$$

Valid for supertankers.

Bulk carriers

Murray (1964–65):

$$W_{St} = 0.026 \cdot L^{1.65}(B + D + T/2) \cdot (0.5 \cdot C_B + 0.4)/0.8$$

Det Norske Veritas (1972):

$$W_{St} = 4.274 \cdot W^{0.62} \cdot L \cdot (1.215 - 0.035 \cdot L/B) \cdot (0.73 + 0.025L/B)$$
$$\cdot (1 + (L - 200)/1800) \cdot (2.42 - 0.07L/D) \cdot (1.146 - 0.0163L/D)$$

W is the section modulus of the midship area. The same limits as for the DNV tanker formula apply, except for $L \leq 380\,\mathrm{m}$.
   More recently, Harvald and Jensen (1992) evaluated cargo ships built in Danish shipyards from 1960 to 1990 with a substantial number built in 1980–1990. The evaluation gives, with 10% accuracy:

$$W_{St} = (L \cdot B \cdot D_A) \cdot C_s$$

$$C_s = C_{so} + 0.064\mathrm{e}^{-(0.5u + 0.1u^{2.45})} \text{ where } u = \log_{10}(\Delta/100\,\mathrm{t})$$

C_{so} [t/m³] depends on ship type:

support vessels	0.0974	bulk carriers	0.0700
tugs	0.0892	tankers	0.0752
cargo ships (1 deck)	0.0700	VLCC	0.0645
cargo ships (2 decks)	0.0760	product carriers	0.0664
cargo ships (3 decks)	0.0820	reefers	0.0609
train ferries	0.0650	passenger ships	0.0580
rescue vessel	0.0232		

Schneekluth's method for dry-cargo ships

The method was developed by Schneekluth (1972). The hull steel weight is first determined for individual section panels which then form the basis for plotting a curve of steel weight per unit length. The advantages over other methods are:

1. Provides a wider range of variation, even for unusual ratios of cargo ship main dimensions.
2. Type of construction of longitudinal framing system is taken into account.
3. Efficient and easy to program.
4. Effect of C_B considered.

Initially, the method was developed for dry-cargo ships by evaluating systematically varied cargo-ship sizes and forms subject to the following boundary conditions:

1. Dry-cargo ships of flush deck construction with bulkheads extending to the topmost continuous deck. The superstructure is assessed in a separate procedure. Hatches are not included.
2. Material strengths, number of bulkheads and height of double bottom in hold area comply with GL regulations of 1967, height of double bottom in machinery space raised by 16%, Class 100A4.
3. Ship form without bulbous bow and rudder heel.
4. Single-screw ship with main engine situated aft; hatchway breadth $\approx$ $0.4B + 1.6$ m, overall length of cargo hatchways $\approx 0.5L$.
5. The following parts of the steel construction are taken directly into account: hatchway structures, engine casing construction, bulwark over 90% of the ship's length, chain locker, chain pipe and chain deck, reinforcements for anchor winch, rudder bearing and shaft tube.
6. Approximately 10% is added to the unit weights to cover the following weights which are not determined individually:
   (a) Increased material scantlings (material management).
   (b) Local reinforcement.
   (c) Heavier construction than prescribed.
   (d) Engine foundations of normal size, masts, posts, rudder body.
   (e) Tank walls in engine room.
7. The following weights are not included in the calculation:
   Hatches
   Special installations (e.g. deep tanks and local strengthening)
   Bulbous bow
   Rudder heel

Essentially, the method takes into account only the following main data:

L	[m]	Length between perpendiculars
L_s	[m]	Length over which sheer extends, $L_s \leq L_{pp}$
B	[m]	Width
D	[m]	Depth to topmost continous deck
T	[m]	Draught at construction waterline
C_B		Block coefficient to construction waterline
C_{BD}		Block coefficient to waterline tangential to topmost continuous deck
C_M		Block coefficient of midship section to construction waterline
s_v	[m]	Height of sheer at forward perpendicular
s_h	[m]	Height of sheer at aft perpendicular
b	[m]	Height of camber of topmost continuous deck at $L/2$
n		number of decks
∇_U	m^3	Volume below topmost continuous deck

In the early design stage, the underdeck volume ∇_U can be approximated as the sum of the hull volume up to the side deck, sheer volume, camber volume and hatchway volume:

$$\nabla_U = \underbrace{L \cdot B \cdot D \cdot C_{BD}}_{\nabla_D} + \underbrace{L_s \cdot B(s_v + s_h)C_2}_{\nabla_s} + \underbrace{L \cdot B \cdot b \cdot C_3}_{\nabla_b} + \underbrace{\sum l_L \cdot b_L \cdot h_L}_{\nabla_L}$$

∇_U is the hull volume to main depth, ∇_s the volume increase through sheer, ∇_b the volume increase through beam camber, and ∇_L the hatchway volumes. The hatchway volumes are the sum of the products of hatchway length l_L, hatchway breadth b_L and hatchway height h_L. A mean value taking account of the camber may be given for h_L.

$$C_2 \approx C_{BD}^{2/3}/6 \approx 1/7$$

$$C_3 \approx 0.7 \cdot C_{BD}$$

$$C_{BD} \approx C_B + C_4 \frac{D-T}{T}(1 - C_B)$$

with $C_4 \approx 0.25$ for ship forms with little frame flare,
$\qquad C_4 \approx 0.4$–0.7 for ship forms with marked frame flare.

These formulae are also useful for other design purposes, since the underdeck volume is important in the early design stage.

The hull steel weight is calculated as the product of the underdeck volume ∇_U, the specific volumetric weight C_1 [t/m³] and various corrective factors:

$$W_{StR} = \nabla_U \cdot C_1$$

$$\cdot \left[1 + 0.033 \left(\frac{L}{D} - 12 \right) \right]$$

$$\cdot \left[1 + 0.06 \left(n - \frac{D}{4m} \right) \right]$$

$$\cdot \left[1 + 0.05 \left(1.85 - \frac{B}{D} \right) \right]$$

$$\cdot \left[1 + 0.2 \left(\frac{T}{D} - 0.85 \right) \right]$$

$$\cdot \left[0.92 + (1 - C_{BD})^2 \right]$$

$$\cdot \left[1 + 0.75 C_{BD} (C_M - 0.98) \right]$$

The formula is applicable for $L/D \geq 9$.

For normal cargo ships ($L = 60$–180 m): $C_1 = 0.103 \cdot [1 + 17$
$$\cdot (L/1000 - 0.11)^2]$$

For passenger ships ($L = 80$–150 m): $C_1 = 0.113$–0.121

For reefers ($L = 100$–150 m): $C_1 = 0.102$–0.116

The formula applies to the ship's hull up to the topmost continuous deck. Hence it also contains a 'continuous superstructure'. Superstructure and deckhouses situated above this limit are treated separately.

Where the superstructure covers most of the ship's length, a depth increased by the height of this superstructure can be used and the ratios L/D, B/D, C_{BD} etc. formed. Next, the volumes not covered by the continuous superstructure must be estimated and subtracted to give the underdeck volume factor ∇_U.

Tankers, bulkers and containerships are better calculated using the earlier mentioned coefficient method.

The cargo decks of ro-ro ships should be designed for high vehicle axle loads and fork-lift operations. This makes them much heavier than usual. Further additional weights are caused by the limits imposed by the web frame depths. The additional weight of ro-ro ships increases in proportion to the width, i.e. the hull steel weight, based on the specifications of a normal dry-cargo vessel, cannot always be corrected using a constant factor.

The result of the hull steel weight equation still has to be corrected for:

1. Bulkhead construction method $+2.5\%\ W_{StR}$
2. Bulbous bow $+0.4$–$0.7\%\ W_{StR}$
   or related to the bulb volume $+0.4\,\text{t/m}^3$

Part of the bulbous bow weight is already included in the calculation result with the underdeck volume.

'Special items' not determined by the steel weight procedure so far include:

Deep tanks: The weight of the additional tank walls is increased by around 30% to account for wall stiffening.

Additional, non-specified bulkheads or specified but not fitted bulkhead (special approval):   Weight of plates plus 40–60% for welded stiffenings, to be calculated from tank top onwards. The vertical variability in the plating is taken into account.

Further amounts may need to be added for special conditions or construction types. The determining factors are:

Bulk cargo, ore: The classification societies require that vessels carrying bulk and ore should be strengthened. Most important is strengthening of the double bottom. This weight should be estimated separately.

Higher double bottom: If the height of the double bottom exceeds GL specifications, the extra steel weight related to the difference in volume between the normal and the raised double floor in longitudinal frames is around $0.1\,\text{t/m}^3$. The following constructional requirements apply here: longitudinal frames, transverse frames only at the narrow ship's ends. Alterations to the upper boom are taken into account here.

Additional steel weight of the higher double bottom: for longitudinal stiffening the volumetric steel weight is around $0.1\,\text{t/m}^3$. For transverse stiffening, the volumetric steel weight is $(0.1 + x/2000)\,\text{t/m}^3$: x is the percentage increase of the double bottom height compared to GL requirements. If, for example, the double bottom is 10% higher than required, $0.105\,\text{t/m}^3$ should be assumed.

Floorplates must be on each frame and side girders 4 m apart.   If side-girders are close together the additional steel weight can increase by one-third. The double bottom volume can be approximated by:

$$\nabla_{db} = L \cdot B \cdot h_{db} \left[C_B - 0.4 \left(1 - \frac{h_{db}}{T} \right)^2 \sqrt{1 - C_B} \right]$$

with h_{db} the absolute height of the double bottom.

Engine foundations: The weight of the engine foundations has already been dealt with in connection with this method for 'normal propulsion systems'. A differential amount must be used for particularly strong plants. Here, 3–6 kg/kW or the following power-related unit weights can be assumed for direct-drive propulsion diesel engines:

$$W_{StF} = \frac{27 P_B}{(n + 250) \cdot (15 + P_B/1000)}$$

where W_{StF} [t] is the weight of the engine foundation, n [min^{-1}] the rpm of the engine, and P_B [kW] the power of the engine.

Container stowing racks: These are discussed in Schneekluth's steel weight calculation for containerships (see below).

Additions for corrosion: If, due to special protective anti-corrosion measures (e.g. coating), additions for corrosion can be disregarded, the steel weight of large tankers will be reduced by 3–5%.

As a very rough estimate, the influence of ice strengthening may be estimated following Dudszus and Danckwardt (1982), Carstens (1967) and N. N. (1975):

Germanischer Lloyd	E	E1	E2	E3	E4		Polar icebreaker
Finnish ice class		IC	IB	IA	IA Super		
Add % in hull steel weight	2	4	8	13	16		Up to 180

The Canadian ice class ranges from Arc 1 to Arc 4. A 180% increase in the hull steel weight can be expected for Arc 4.

Reducing weight by using higher tensile steel

Higher tensile steel has roughly the same modulus of elasticity as mild ship-building steel. For this reason, buckling strength and vibration behaviour of structures should be carefully considered when using higher strength steels instead of mild steel. Use of high tensile steel in bottom and deck can reduce weight by 5–7%.

Schneekluth method for containerships

The method (Schneekluth, 1985) is based on the evaluation of systematically varied ship forms and sizes of a containership type corresponding to the level of development at the early 1980s. To isolate the influence of the main data and ratios on the hull steel weight, the construction and building method was kept as uniform as possible over the entire variation range. Checked using statistical investigations, this corresponds reasonably consistently to practical reality and the building method applied in shipyard. The following boundary conditions for the method result:

(1) General data on type and construction

1. Full scantling vessel with freeboard in open double-hull construction, i.e. with broad hatchways and longitudinal bulkheads below the longitudinal hatchway coamings.
2. The bulkhead spacings and number of bulkheads are adapted to those of conventional containerships.
3. The forecastle has an average length 10% L, including its extension which embraces the forward hatchway on both sides.
4. The forecastle height is 2.7 m throughout.
5. Unlike the method described for normal cargo ships, the forecastle steel weight is taken into account directly with the hull steel weight. Correspondingly the forecastle volume is calculated as part of the underdeck volume ∇_U. As in the method used for cargo ships, other superstructure and deckhouses are calculated separately.
6. The hatchway length (i.e. the sum of the aperture lengths) is 0.61–0.65 L.
7. The hatchway coaming height is 0.8–1.3 m.
8. The length of the hatchway area between the foremost and aftmost end coamings is 0.72–0.74L. Where the ship's length is great, the hatchway area consists of two sections forward and aft of the engine room.
9. The hatchway widths are taken to be restricted, as is usual owing to the pontoon hatch cover weights. On the smallest ships, these are restricted to five container widths (approximately 13.5 m), on the larger ships to four container widths (approximately 10.5 m). Where six and eight containers are positioned adjacently near amidships, allowance is made for a longitudinal web between the hatchways. Where ships have seven, nine and ten adjacent containers, two longitudinal webs are assumed.
10. Irrespective of the dimensional pattern of container stowage, the main dimensions L, B, D can also be considered continuously variable on containerships. The apparent inconsistency is particularly noticeable for the width. The statistics of existing ships show, however, that the normal variation range of the side-tank breadth produces the variability required to assume a continuous change in width. On this basis the method starts with an average side-tank breadth of 2.25 m.

(2) Form, speed, propulsion

1. Single-screw vessel with bulbous bow and without rudder heel.
2. Diesel propulsion with a typical value for the propulsion power of around 0.6 kW/t displacement. $F_n < 0.26$.
3. $0.52 \leq C_B \leq 0.716$.
4. In ships of short or medium length the engine room lies aft and has a length of 14–15% L. In ships exceeding 200 m in length the engine room lies forward of the last hold and has a length of 12–13% L.
5. A normal midship section form will be used.

(3) Construction and strength

1. Standard building method with longitudinal frames in the upper and lower booms and with transverse frames in the side-walls and at the ship's ends.
2. Material strengths in accordance with GL 1980, Class 100 A4, without ice strengthening. According to the speed range established, bottom reinforcement in the foreship will only be used in the normal, not in the extended, area.
3. Double bottom height in hold area and in engine room generally 16% higher than GL minimum. Stepping-down of double bottom at forward end as usual, corresponding to container stowage.
4. Transverse and longitudinal cross-bars between the hatchways are enlarged to form box beams and are supported at points of intersection. Longitudinal hatchway coamings extend downwards into the wing tank side.
5. The section modulus is 10% above the normal minimum value as due to the open design torsional strength has to be considered in addition to the usual longitudinal strength.

The upper section of the wing tank at a height of 2.4 m is assumed to be of higher strength steel HF 36 between engine room and forecastle. On ships over 200 m in length the floor of the gangway, which forms the upper part of the wing tank, also consists of high-tensile (HT) steel. While HT steel is rarely used in the upper decks of smaller ships except for the hatch coamings, in this weight estimation procedure it is considered (in terms of weight) the norm for all ship sizes. HT steel is generally more economical and conventional for containerships longer than 130 m. For all ships, the frame spacing beyond the ship's ends amounts to:

Transverse framing	750–860 mm
Longitudinal framing, bottom	895 mm
Longitudinal beams	750 mm

This frame spacing is more than sufficient for the short variants below a length of 130 m. Frame spacing adapted to ship length may produce weight savings of about 5% for shorter ships.

(4) Dimensional constraints

The method can be applied to ships 100–250 m in length and for widths including the Panama maximum width of 32.24 m. The main ratios have been varied within the following limits:

L/B from 7.63 to 4.7, with small ships to 4.0

L/D from 15.48 to 8.12

B/D from 1.47 to 2.38

B/T from 2.4 to 3.9 with $T = 0.61D$ and

from 1.84 to 2.98 with $T = 0.8D$

C_B from 0.52 to 0.716.

Extrapolation beyond these limits is possible to a certain extent.

(5) Steel weights determined in the formula

The following components and factors are taken into account:

1. Forecastle of the above-mentioned standard dimensions.
2. Bulbous bow.
3. 'Tween decks in the engine room and hold area (gangway in upper section of wing tank)
4. Top plates and longitudinal supports of the main engine foundations.
5. Hatchway coamings (if not extreme in height), chain lockers.
6. Chain pipes and chain deck pipes.
7. Increased material strengths (from stock).
8. Deposited metal.
9. Bracket plates, minor items and small local reinforcement.
10. Masts, posts.
11. Rudder structure.
12. Local strengthening of inner bottom. This assumes that the side supports roughly fit the corners of the container stack.

Not included or determined in the formulae are:

1. Hatch covers.
2. Container cell guides.
3. Ice-strengthening.
4. Speed or performance-conditioned strengthening such as above average bottom reinforcement in the forebody.
5. Rudder heel.
6. Special installations and local strengthening.
7. Construction types more expensive than regulation, apart from the above-mentioned 10% increase in midship section strength modulus.

The input values for the method are virtually the same as those used for a normal cargo ship, except for:

1. The deck number is always 1.
2. The forecastle volume is included in the underdeck volume ∇_U.

The following equation should be used to calculate the hull steel weight of containerships:

$$W_{StR} = \nabla_U \cdot 0.093$$
$$\cdot \left[1 + 2(L - 120)^2 \cdot 10^{-6} \right]$$
$$\cdot \left[1 + 0.057 \left(\frac{L}{D} - 12 \right) \right]$$
$$\cdot \sqrt{\frac{30}{D + 14}}$$
$$\cdot \left[1 + 0.1 \left(\frac{B}{D} - 2.1 \right)^2 \right]$$

$$\cdot \left[1 + 0.2 \left(\frac{T}{D} - 0.85 \right) \right]$$

$$\cdot \left[0.92 + (1 - C_{BD})^2 \right]$$

Depending on the steel construction the tolerance margin of the result will be somewhat greater than that of normal cargo ships. The factor before the first bracket may vary between 0.09 and 0.10.

The formula is similar to that for normal cargo ships except:

1. The underdeck volume ∇_U contains the volume of a short forecastle and the hatchways.
2. $L/D \geq 10$.

Further corrections:

1. Where normal steel is used the following should be added:

$$\Delta W_{St} = 0.035(\sqrt{L} - 10) \cdot \left[1 + 0.1 \left(\frac{L}{D} - 12 \right) \right] W_{StR}$$

   This correction applies to ships between 100 m and 180 m in length. One of the reasons for this addition—relatively large for long ships—is that both the high material strengths in the deck and those of the side-walls must be arranged stepwise.
2. No correction for the wing tank width is needed. The influence is slight.
3. This formula can also be applied to containerships with trapezoidal midship sections. These are around 5% lighter.
4. As in the method for normal cargo ships, further corrections can be added, i.e. for ice-strengthening, different double bottom height, higher speed and higher hatchways.
5. The unit weight of double bottoms built higher than stipulated by GL amounts to $40 + x/2$ [kg/m^3] when related to the hold difference. Here x is the percentage increase in double bottom height relative to the required minimum by GL. This formula applies to longitudinal frame construction with transverse framing at the ends of the ships and widely spaced longitudinal girders.

Container cell guides are often included in the steel weight, while lashing material and 'cooling bars' are considered to be part of the equipment.

Weights of container cell guides:

Container Type	Length (ft)	Fixed	Detachable
Normal	20	0.70 t/TEU	1.0 t/TEU
Normal	40	0.45 t/TEU	0.7 t/TEU
Refrigerated	20	0.75 t/TEU	—
Refrigerated	40	0.48 t/TEU	—
Refrigeration	20	0.75 t/TEU	
Pipes	40	0.65 t/TEU	

Where containers are stowed in three stacks, the lashings weigh:

For 20 ft containers   0.024 t/TEU

For 40 ft containers   0.031 t/TEU

For mixed stowage   0.043 t/TEU

Position of hull steel centre of mass

The height of the hull steel centre of weight, disregarding superstructure and deckhouses, is largely independent of ship type and can be determined by:

$$\overline{KG}_{StR} = \left[58.3 - 0.157 \cdot (0.824 - C_{BD}) \cdot \left(\frac{L}{D}\right)^2\right] D_s \cdot 0.01 \approx 0.057 D_s$$

D_s is the depth increased to take account of the sheer and the hatchway volume. The correction for the sheer could be calculated if the sheer continues to around midships. The formula values can be corrected as follows:

For bulbous bow $- 0.004\ D$

For L/B differing from $L/B = 6.5$ $+ 0.008D$ for $\Delta L/B = \pm1.0$

If a set of hydrostatic curves is available the steel centre of weight can also be calculated as a function of the height of the sectional area centre of weight (including the hatchways). The hull steel centre of weight is then some 5% below the centroid of the enclosed volume. The value can be corrected as with the formula given above.

The longitudinal position of the hull steel centre of weight lies

1. approximately at the centre of volume of the capacity curve; or
2. half-way between the forward perpendicular and the aft edge of the main deck.

Weights of superstructure and deckhouses

The method (Müller-Köster, 1973) is based on the requirements of the classification societies. Scantlings for superstructure and deckhouses are commonly bigger than specified for reasons of production. Therefore, it is recommended to add a further 10%–25% to the result of the calculation.

Forecastle

The volumetric weight of a forecastle is:

For ships with $L \geq 140$ m: $C_{\text{forecastle}} \approx 0.1$ t/m^3

For ships with $L \approx 120$ m: $C_{\text{forecastle}} \approx 0.13$ t/m^3

The values apply to a forecastle height of 2.5–3.25 m and a length of up to 20% L_{pp}.

A forecastle of around one-third L_{pp} in length causes a 10% decrease in value. If the height of the forecastle extends over two decks, the volumetric weight can be expected to decrease by 5–10%.

Poop

The volumetric weight of a poop which extends to the forwardmost engine room bulkhead of an engine room located aft is $C_{\text{poop}} = 0.075\,\text{t/m}^3$. A long poop which covers one hold in addition to the engine room is around $0.09\,\text{t/m}^3$.

Deckhouses

Usually the material scantlings of deckhouses are reinforced beyond the requirements of classification societies. This is because:

1. It reduces aligning and straightening out during building.
2. It strengthens the material against corrosion—especially in the lower area.
3. The greater distance between stiffeners means less welding.

These additions are partly included in the method. It is recommended, however, to add 15% to the following values for winch houses and 7–10% for other deckhouses. The large amounts added for winch houses include the U supports fixed to the deck as foundations for the winches.

Houses with living quarters

Deckhouses extending over several decks are not regarded as one complex but taken in sections and placed in order of position above the upper deck. Thus in Table 5.2 the deckhouse section situated on the topmost continuous deck is called layer I, the one above this, layer II, etc. So a deckhouse situated on a poop starts with layer II.

Table 5.2 Volumetric deckhouse weight C_{DH} [t/m^3] as a function of the area relationship F_o/F_u

			Layer		
F_o/F_u	*I*	*II*	*III*	*IV*	*Wheelhouse*
1.0	0.057	0.056	0.052	0.053	0.040
1.25	0.064	0.063	0.059	0.060	0.045
1.5	0.071	0.070	0.065	0.066	0.050
1.75	0.078	0.077	0.072	0.073	0.055
2.0	0.086	0.084	0.078	0.080	0.060
2.25	0.093	0.091	0.085	0.086	0.065
2.5	0.100	0.098	0.091	0.093	0.070

The weight depends on: construction form, number of decks, length of ship, deck height, length of internal walls and the ratio of the upper deck area F_0 and outside walkway to the area actually built over F_u. Table 5.2 shows the volumetric weight of the individual layers (Fig. 5.3).

The weight of one deckhouse section is given by:

$$G_{DH} = C_{DH} \cdot F_u \cdot h \cdot K_1 \cdot K_2 \cdot K_3$$

C_{DH} [t/m^3] from the table, interpolated for intermediate values of F_o/F_u
h is the deck height
Correction K_1 for non-standard deck height: $K_1 = 1 + 0.02(h - 2.6)$

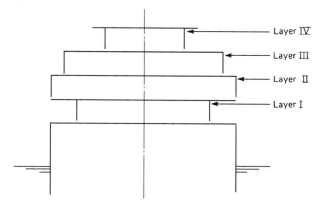

Figure 5.3 Breaking down the deckhouse weight into individual layers

Correction K_2 for non-standard internal walls (which is 4.5 times the
   deckhouse section length):
$K_2 = 1 + 0.05(4.5 - f_i)$ with f_i = internal wall length/deckhouse section
   length
Correction K_3 for ship length: $K_3 = 1 + (L_{pp} - 150) \cdot 0.15/130$ for
   $100\,\text{m} \le L_{pp} \le 230\,\text{m}$

Müller-Köster (1973) gives special data for winch houses.

Taking local stiffening below the winch house and the winch foundations
themselves into account can make the winch houses considerably heavier. A
70% addition is recommended here.

The height of the centre of weight for superstructure and deckhouses (in
relation to the deck height h in each case) is calculated separately for each
deck. It is around 0.76–$0.82h$ if no internal walls exist and $0.7h$ otherwise.

Using light metal

Owing to the danger of corrosion, only light metal alloys without copper,
usually aluminium–magnesium, should be used to save weight. An aluminium
superstructure or deckhouse must be insulated on the steel hull side, e.g. by
putting riveted joints with plastic insulation strips between the plates and small
plastic tubes between the rivets and the walls of the rivet holes or by using
explosion plated elements.

Since aluminium alloys have a comparatively low melting point, fire protection has to be provided by proper insulation.

The possible weight savings are often over-estimated. The light metal
weights of superstructure, deckhouses and possibly other special installations
can be assumed to be 45–50% of the steel weight.

Deckhouses made of lightweight metal cost about 5–7 times the amount of
steel deckhouses. It is not only the metal itself which is more expensive than
the steel, but also its processing, since most steel processing machines are not
designed to work light metal. Welding light metal is also more costly. Hull
components made of light metal are often manufactured by specialized firms.

5.2 Weight of 'equipment and outfit' (E&O)

Because ships have increased comfort, weight of E&O has increased. Despite smaller crews, the weight of outfit has increased because:

1. Greater surface area and space required per man.
2. The incombustible cabin and corridor walls in use today are heavier than the earlier wooden walls.
3. Sanitary installations are more extensive.
4. Air-conditioning systems are heavier than the simple ventilation devices formerly used.
5. Heat and vibration insulation is now installed.

The weight of some equipment items has increased over time:

1. The weight of hatches:
   (a) Owing to the application of steel in the lower decks.
   (b) Owing to greater hatchway areas.
   (c) Owing to the requirement for container stowage on the hatches.
2. More comprehensive cargo gear.
3. Fire prevention measures (CO_2 units and fire-proof insulation).

In contrast, the hold ceiling is now lighter. Nowadays the side-ceiling of holds is normally omitted and instead of the bottom ceiling it is usually the actual inner bottom which is strengthened. This strengthening is included in the steel weight.

Two methods for subdividing the E&O components are commonly applied:

1. According to the workshops and the company departments which carry out the work.
2. According to the function of the components and component groups.

These or similar component subdivision methods—extended to cover machinery—provide a detailed and comprehensive basis for the whole operation (calculation, construction, preparation, procurement of material) at the shipyard.

Details of a ship's lightweight and its subdivision are rarely published. Neither is there a method to determine the weight of E&O. If no reliable data on the basis ship exists, published statistical values have to be used. These values may relate to a variety of component and ship sizes. What proportion of the ship's lightweight is made up by E&O depends to a great extent on the ship type and size.

Better estimates of E&O weights may be obtained if E&O is divided into general E&O and cargo-specific E&O. The shipyard can use larger databases to derive empirical estimates for the general E&O.

An exacter knowledge of the E&O weights can only be gained by breaking down the weight groups and determining each weight individually. This involves gathering information from the subcontractors. As this procedure is rather tedious, the degree of uncertainty for these weight groups remains generally larger than for steel weight.

The following are the main methods used to determine E&O weights:

1. The construction details are determined and then the individual weights summed. This also enables the centre of weight of this weight group to be ascertained. Furthermore, the method provides a sound basis for the calculation. This very precise method requires a lot of work. It is therefore unsuitable for project work. A comprehensive collection of unit weights for the construction details is also necessary.
2. The sum total of all E&O weights is determined by multiplying an empirical coefficient with a known or easily obtainable reference value. This method of attaining a combined determination of all E&O weights will produce sufficiently precise results only if data for well-evaluated 'similar ships' exist. Nevertheless, this method is by far the most simple. If no suitable basis ships and their data are available, the coefficients given in the literature can be used.

   The coefficients depend on the ship type, standard of equipment and ship size. Where possible, the coefficients should be related to ship's data which produce a more or less constant value for the ship's size. The coefficient then depends only on ship type and standard of equipment.

On all types of cargo ships, the equipment weight increases approximately with the square of the linear dimensions. Reference values here are primarily area values, e.g. $L \cdot B$ or the 2/3 power of volumes. On passenger ships, however, the equipment weight increases approximately with the 'converted volume'. Particularly suitable reference values are:

1. The 'converted volume'—including superstructure and deckhouses corresponding to the gross volume of tonnage measurement of 1982.
2. The steel weight.

Literature on the subject gives the following reference values:

1. The 'converted volume' $L \cdot B \cdot D$ (Henschke, 1965).
2. The area $(L \cdot B \cdot D_A)^{2/3}$. Here, D_A is 'depth-corrected to include the superstructure', i.e. the normal depth D increased by an amount equal to the superstructure volume divided by the deck area. The values scatter less in this case than for (1) (Henschke, 1952).
3. The area $L \cdot B$. Here, too, the values are less scattered than for the reference value $L \cdot B \cdot D$. Weberling (1963) for cargo ships, Weberling (1965) for tankers and reefers, Watson and Gilfillan (1977).
4. The steel weight W_{St}.
5. The hold volume. Krause and Danckwardt (1965) consider not only summary weights, but also individual contributions to this weight group.
6. The hold volume associated with the deadweight.

E&O weights for various ships

Passenger ships—Cabin ships

$$W_o = K \cdot \sum \nabla$$

Here, $\sum \nabla$ is the total 'converted volume' and $K = 0.036$–$0.039 \, \text{t/m}^3$.

Passenger ships with large car-transporting sections and passenger ships carrying deck passengers

Formula as above, but $K = 0.04\text{–}0.05\,\text{t/m}^3$.

Cargo ships of every type

$$W_o = K \cdot L \cdot B$$

Cargo ships	$K = 0.40\text{–}0.45\,\text{t/m}^2$
Containerships	$K = 0.34\text{–}0.38\,\text{t/m}^2$
Bulk carriers without cranes:	
with length of around 140 m	$K = 0.22\text{–}0.25\,\text{t/m}^2$
with length of around 250 m	$K = 0.17\text{–}0.18\,\text{t/m}^2$
Crude oil tankers:	
with length of around 150 m	$K \approx 0.28\,\text{t/m}^2$
with length of over 300 m	$K \approx 0.17\,\text{t/m}^2$

Henschke (1965) gives summary values for E&O weight on dry-cargo ships and coastal motor vessels:

$$W_o = \frac{0.07 \cdot (2.4 - \nabla_{LR}/W_{dw})^3 + 0.15}{-1 + \log \nabla_{LR}} \cdot \nabla_{LR}$$

$$\nabla_{LR} = \text{hold volume [m}^3\text{]}$$
$$\nabla_{LR}/W_{dw} = \text{stowage coefficient [m}^3\text{/t]}$$

The formula is applicable in the range $1.2 < \nabla_{LR}/W_{dw} < 2.4$.

The traditional statement that in dry-cargo ships the E&O weight roughly equals the weight of the engine plant, is no longer valid. The E&O weight is frequently 1.5–2 times that of the engine plant.

Reefers (between 90 m and 165 m in length)

$W_o = 0.055L^2 + 1.63\nabla_i^{2/3}$ where $L = L_{pp}$ and ∇_i is the gross volume of insulated holds. The formula is based on the specifications of reefers built in the 1960s (Carreyette, 1978).

Application of a special group subdivision to determine E&O weights

This method entails considerably less work than the precise, but complicated, process of establishing the weights of each construction detail. On the other hand, it is more precise and reliable than determining the overall E&O weight using only one coefficient.

The individual components are classified according to how they are determined in the calculation and their relationship to type, rather than using aspects of production and function. Four groups are formed and the precise weight breakdown and data of each given. The method is applicable primarily to bale cargo ships and containerships, and has the added benefit of facilitating the estimation of the centre of weight. Modified correctly, the method can also be applied to other ship types.

By putting individual weights into the calculation, the differences in ships of similar size and function which have varying standards of equipment can be partly taken into account. Although some of the less easily calculated weights can still only be ascertained using a coefficient, the degree of variation in the overall result is reduced. There are three main reasons for this:

1. Large individual weights are more precisely known and no longer need to be estimated.
2. Coefficients are used only where there are relatively authorative reference values (e.g. outfit areas) or where the components to be determined are largely independent of the ship type. This diminishes the risk of error.
3. If there are several individual weights to calculate, it is highly probable that not all the errors will have the same sign. Even though the individual estimations or individual coefficients are no more precise than an overall coefficient for the overall weights group, errors with opposing signs will usually partly cancel each other.

The following component weights groups are used in the method of Schneekluth.

Group I Hatchway covers

Ship-type dependent; individual weights, relatively easily determined given at least approximate knowledge of hatchway size from contractor specifications or using empirical values.

Group II Cargo-handling/access equipment

Highly dependent on ship type or largely pre-determined for the individual design. Calculated from a limited number of individual weights.

Group III Living quarters

Limited dependence on type; can be determined relatively precisely using coefficients, since the 'converted' volume or the surface area of the living quarters provide authorative reference values.

Group IV Miscellaneous

Comprises various components which are difficult to calculate individually, but relatively independent of ship type and thus can be determined using a ship-size-related coefficient.

Breakdown of weight group E&O with reference values to determine sub-group weights

Group I Hatchway covers

Group I comprises all cargo hatches and any internal drive mechanisms—but no exterior drive mechanisms.

Weather deck—'single pull' with low line system

	Weight in kg/m hatchway length				
Hatchway breadth [m]	6	8	10	12	14
Load 1.75 t/m²	826	1230	1720	2360	3150
1 container layer	826	1230	1720	2360	3150
2 container layers	945	1440	2010	2700	3550

The 20 ft/20 t containers are calculated as evenly distributed over the length. In the 'Piggy Back' system, the weights mentioned above are around 4% less.

The hatchway cover weight can be approximated using a formula given by Malzahn. The weight of single-pull covers on the weather deck carrying a load of 1.75 t/m² is

$$W_l/l = 0.0533 \cdot d^{1.53}$$

where W_l is the cover weight [t], l the cover length [m] and $d =$ the cover breadth [m].

Tween deck—folding cover design—not watertight

The covers are 0.2 m broader than the clear deck opening.

	Weight in kg/m hatchway length				
Cover breadth [m]	6	8	10	12	14
Normal load	845	1290	1800	2440	3200
Fork-lift operation	900	1350	1870	2540	3360
2 container layers	930	1390	1940	2600	3460

Using GL specifications, the normal load applies to a deck height of 3.5 m. The fork-lift trucks have double pneumatic tyres and a total weight of 5 t. The container layers consist of 20 ft containers with a 20 t evenly distributed load.

Pontoon covers are lighter (up to around 15%).

Group II Loading equipment

Derricks, winches, cargo gear, deck cranes, hold ceiling, container lashing units—excluding king posts which are classified under steel weight.

Light cargo derricks

Fabarius (1963) gives derrick weights. These are not discussed here as modern general cargo ships are usually equipped with cranes instead.

Winches used for handling cargo

The weight of cargo-handling winches depends on their lifting capacity, lifting speed and make or construction type (Ehmsen, 1963). Where no published data are available, a weight of 0.6–1 t per ton lifting capacity should be assumed for simple winches. In terms of their pull, winches for derricks with lifting capacity varying according to rigging of cargo cables are designed to the lower value, e.g. 3 t for a 3–5 t boom, but have a higher rpm rate for the higher value. They weigh 1–2 t per ton lifting capacity, in relation to the lower lifting value. The other winches—hanger winch, preventer winches, belly guy winches and the control console—weigh roughly the same for one boom pair as the two loading winches together.

Deck cranes installed on board

If manufacturers' data are not available, the dimensions and weights of ships' cranes can be taken from the following table:
    Weights for deck cranes installed on board:

Max.	Weight (t) at max. working radius			
load (t)	15 m	20 m	25 m	30 m
10	18	22	26	
15	24	28	34	
20		32	38	45
25		38	44	54
30		42	48	57
35		46	52	63

The height of the centre of weight of the crane in the stowed position (with horizontal jib) is around 20–35% of the construction height, the greater construction heights tending more towards the lower percentage and vice versa. The heights are measured relative to the mounting plate (i.e. to the upper edge of the post).

Inner ceiling of hold

The holds of bale-cargo vessels are rarely fitted with a ceiling (inner planks) today. The extent of the ceiling is either specified by the shipping company or a value typical for the route is used. Should a ceiling be required, its weight is easy to determine. The equivalent volume of wood in the hold—projected area of hold sides multiplied by 0.05 m thickness—can be used for side-planking with lattices. The bulkhead ceilings and 10% of the wood weight for supports must be added to this. Pine wood is normally used for the floor ceiling. Longitudinal layers of planks 0.08 m thick are secured to 0.04 m × 0.08 m transverse battens arranged above each frame. In the absence of floor ceilings, the steel plate thickness has to be increased, especially in bulk carriers subjected to loads from grab discharge.

Group III Accommodation

The E&O in the living quarters include:

Cabin and corridor walls—if not classed as steel weight.
Deck covering, wall and deck ceiling with insulation.
Sanitary installations and associated pipes.
Doors, windows, portholes.
Heating, ventilation, air-conditioning and associated pipes and trunking.

Kitchens, household and steward's inventory.
Furniture, accommodation inventory.

All weights appertaining to the accommodation area can be related to the surface area or the associated volume. The engine casing is not included in the following specifications.

The specific volumetric and area weights increase to some extent with the standard of the facilities, the ship's size and the accommodation area, since a larger accommodation area means an increase in pipe weights of every type. The greater volumes typical of larger ships have an effect on the specific weights per unit area. The specific volumetric and unit area weights are:

For small and medium-sized cargo ships: 160–170 kg/m^2 or 60–70 kg/m^3

For large cargo ships, large tankers, etc.: 180–200 kg/m^2 or 80–90 kg/m^3

Group IV Miscellaneous

Group IV comprises the following:

Anchors, chains, hawsers.
Anchor-handling and mooring winches, chocks, bollards, hawse pipes.
Steering gear, wheelhouse console, control console (excluding rudder body).
Refrigeration plant.
Protection, deck covering outside accommodation area.
Davits, boats and life rafts plus mountings.
Railings, gangway ladders, stairs, ladders, doors (outside accommodation area), manhole covers.
Awning supports, tarpaulins.
Fire-fighting equipment, CO_2 systems, fire-proofing.
Pipes, valves and sounding equipment (outside the engine room and accommodation area).
Hold ventilation system.
Nautical devices and electronic apparatus, signaling systems.
Boatswain's inventory.

Weight group IV is primarily a function of the ship's size. There is a less marked dependence on ship type. The weight of this group can be approximated by one of the following formulae:

$$W_{IV} = (L \cdot B \cdot D)^{2/3} \cdot C \quad 0.18\,\text{t/m}^2 < C < 0.26\,\text{t/m}^2$$
$$W_{IV} = W_{St}^{2/3} \cdot C \quad 1\,\text{t}^{1/3} < C < 1.2\,\text{t}^{1/3}$$

Other groups: For special ships, parts II and IV may be treated separately, e.g. hold insulation and refrigeration in reefers or pipes in tankers.

Centres of mass of E&O

1. If the weights of component details are given, their mass centres can be calculated or estimated. A moment calculation then determines the centre

of mass of the group. Determining details of the weight is advantageous in evaluating:

(a) Weight.

(b) Centre of mass.

(c) Price.

2. If the weight is determined initially as a total, this can be divided up into groups. After estimating the group centres of mass, a moment calculation can be conducted.

3. Using the centres of mass of similar ships—for the whole area of E&O. Typical values of the overall centre of mass are:

$$\text{Dry-cargo ships } \overline{KG}_{MO} = 1.00 - 1.05 D_A$$

$$\text{Tankers } \overline{KG}_{MO} = 1.02 - 1.08 D_A$$

D_A is the depth increased by the ratio superstructure volume divided by the main deck area, i.e. the depth is corrected to include the superstructure by increasing the normal depth by the height of the layer produced by distributing the volume of the superstructure on the main deck.

5.3 Weight of engine plant

The following components and units form the weight of the engine plant:

1. The propulsion unit itself, consisting of engines with and without gearboxes or of a turbine system incorporating steam boilers.
2. The exhaust system.
3. The propellers and energy transmission system incorporating shaft, gearbox, shaft mountings, thrust bearing, stern gland.
4. The electric generators, the cables to the switchboards and the switchboards themselves.
5. Pumps, compressors, separators.
6. Pipes belonging to the engine plant, with fillings. This includes all engine room pipes with filling and bilge pipes located in the double bottom for fuel and ballast.
7. Evaporator and distilling apparatus.
8. Disposal units for effluents and waste.
9. Special mechanisms such as cargo refrigeration and, in tankers, the cargo pump systems.
10. Gratings, floor plates, ladders, sound, vibration and thermal insulation in the engine room.

Criteria for selection of the propulsion system

Choice of the propulsion power is arbitrary. However, it must be sufficient for manoeuvring. The choice of the main propulsion unit is influenced to some extent by the weight of the unit, or the sum of the weights of propulsion unit and fuel. This is particularly the case with fast ships, where the installed weight has a considerable bearing on economic efficiency. In diesel engine drive, the upper power limit is also important. If the power requirement exceeds this limit, one of the following can be applied:

1. Several diesel engines via a gearbox.
2. Multi-screw propulsion with direct-drive diesels or gearbox.
3. Gasturbines via a gearbox.

According to Protz (1965), the following criteria are important in the choice of the propulsion unit:

1. Initial costs.
2. Functional reliability.
3. Weight.
4. Spatial requirements.
5. Fuel consumption.
6. Fuel type.
7. Maintenance costs.
8. Serviceability.
9. Manoeuvrability.
10. Noise and vibration.
11. Controllability.

Ways of determining weight

The weights of the complete engine plant can be determined using the following methods:

1. Using known individual component weights.
2. Using unit weights from similar complete plants.
3. As a function of the known main engine weight.
4. Using weight groups which are easy to determine plus residual weight group.

Using individual weights

Here the weight of water and oil in pipes, boilers and collecting tanks is part of the engine plant weight. The weights of all engine room installations and small components should also be determined.

Engine plant weight using unit weights

If the weight of the plant is established using unit weights of similar complete plants, these will contain specifications for each detail of the engine plant—even the electrical unit, although there is no direct connection between the weight of the propulsion unit and the electrical unit. Ideally the weights of propulsion and electric unit should be treated separately. If the unit weights of existing ships are used as a basis, these should always be related to the nominal power (100%). The designs of the systems should be similar in the following respects:

1. Type of propulsion unit (diesel engine, steam turbine, gas turbine).
2. Type of construction (series engine, V-type engine, steam pressure).
3. RPM of propulsion unit and propeller.
4. Size of ship and engine room.
5. Propulsion power.
6. Auxiliary power.

Given these conditions, unit weights, often ranging from below 0.1 to above 0.2 t/kW, give reliable estimates (Krause and Danckwardt, 1965; Ehmsen, 1974a,b).

Determination of engine plant weights from main engine weights (for diesel units)

Although the determination of the weight of the engine plant as a function of a known main engine weight is in itself a rather imprecise method, it will nevertheless produce good results if basis ship data are available. In the absence of manufacturers' specifications, the following values relating to a 'dry' engine (without cooling water and lubricant) can be used as approximate unit weights for diesel engines:

slow-speed engines (110–140 rpm)	0.016–0.045 t/kW
medium-speed engines in series (400–500 rpm)	0.012–0.020 t/kW
medium-speed V-type engines (400–500 rpm)	0.008–0.015 t/kW

C is also around the upper limit where ships have substantial additional machinery weight (classed as part of the engine plant), e.g. tankers, reefers.

Gearbox weights

Gearbox weights are based on catalogue specifications. Factors influencing the weight include power, thrust, speed input and output, the basic design, i.e. integral gearbox, planetary gearbox, and whether the gearbox is cast or welded. For welded single-reduction and integral gearboxes giving a propeller speed of 100 rpm, a power-related weight of 0.003–0.005 t/kW can be assumed. Where propeller speeds n are not fixed, values can be chosen within the normal limits:

$$W_{Getr} = 0.34\text{–}0.4\frac{P_B}{n}\frac{\text{t}\cdot\text{rpm}}{\text{kW}}$$

Cast gearboxes are approximately three times as heavy.

The use of weight groups to determine engine plant weight

Using easily determined weight groups to calculate engine plant weight is primarily suitable for diesel units. The weight of the unit can be divided up as follows:

1. Propulsion unit
   Engine— using catalogue or unit weight
   Gearbox— using catalogue or unit weight
   Shafting— (without bearing) using classification length and diameter
   For material with tensile strength 700 N/mm², the diameter of the shaft end piece is:
   $d = 11.5(P_D/n)^{1/3}$ d in cm, P_D in kW, n in rpm
   The associated weight is: M/l [t/m] $= 0.081(P_D/n)^{2/3}$
   Propeller— A spare propeller may have to be taken into account. The following formula can be used for normal manganese bronze propellers:
   $W_{Prop} = D^3 \cdot K(\text{t/m}^3)$

for fixed-pitch propellers: $K \approx 0.18 A_E/A_0 - (Z-2)/100$ or
$$K \approx (d_s/D) \cdot (1.85 A_E/A_0$$
$$-(Z-2)/100)$$
with d_s the shaft diameter
Controllable-pitch propellers for cargo ships $K \approx 0.12{-}0.14$
Controllable-pitch propellers for warships $K \approx 0.21{-}0.25$
Ehmsen (1970) gives weights of controllable-pitch propellers. Fixed-pitch propellers on inland vessels are usually heavier than the formulae indicate—the same applies to ice-strengthened propellers and cast-iron spare propellers.

2. Electrical units
Generators powered by diesel engines operate via direct-drive at the same speed as the engines. For turbo-generators, the turbine speed is reduced by a gearbox to a speed matching the generator characteristics. The shaft generator arrangement (i.e. coupling of generator to the main propulsion system) has the following advantages over an electricity-producing system which incorporates special propulsion units for the generators:

1. The electricity is produced by the more efficient main engine.
2. The normally cheaper fuel oil of the main engine is used to produce the electricity.
3. There is no need for any special servicing or repair work to maintain the generator drive.

The use of a shaft generator often requires constant engine speed. This is only compatible with the rest of the on-board operation if controllable-pitch propellers are used and the steaming distance is not too short. Separate electricity producing units must be installed for port activity and reserve requirements.

In the weight calculation, the electrical unit weight includes the generators and drive engines, usually mounted on the same base. Switchboards and electric cables inside the engine room are determined as part of 'other weights' belonging to the engine plant weight. The weight of diesel units is:

$$W_{Agg} = 0.001 \cdot \frac{P}{\text{kW}} \cdot \left(15 + 0.014 \cdot \frac{P}{\text{kW}} \right)$$

The output of the individual unit, not the overall generator output, should be entered in this formula (Wangerin, 1954).

There are two ways to determine the amount of electricity which generators need to produce:

1. Take the sum of the electrical requirements and multiply this with an empirical 'simultaneity factor'. Check whether there is enough power for the most important consuming units, which in certain operational conditions have to function simultaneously.
2. Determine directly using statistical data (Schreiber, 1977).

3. Other weights
Pumps, pipes, sound absorbers, cables, distributors, replacement parts, stairs, platforms, gratings, daily service tanks, air containers, compressors, degreasers, oil cooler, cooling water system, control equipment, control room, heat and sound insulation in the engine room, water and fuel in

pipes, engines and boilers. This weight group is a function of the propulsion power, size of ship and engine room and standard insulation. As a rough estimate:

$$M = 0.04\text{–}0.07P \, \text{t/kW}$$

The lower values are for large units of over 10 MW.
4. Special weights—on special ships

1. Cargo pumps and bulk oil pipes.
2. Cargo refrigerating system (including air-cooling system without air ducts). Weights of around 0.0003 t/(kJ/h) or 0.014 t/m³ net net volume.

The refrigeration system on refrigeration containerships weighs ~1 t/40 ft container with brine cooling system, ~0.7 t/40 ft container with direct vaporization. The air ducts on refrigeration containerships weigh ~0.8 t/40 ft container with brine cooling system, ~1.37 t/40 ft container with direct vaporization. The insulation is part of weight group 'E&O'. Its weight, including gratings and bins, is ~0.05–0.06 t/m³ net net hold volume, or 1.9 t/40 ft container when transporting bananas, 1.8 t/20 ft container when transporting meat.

Propulsion units with electric power transmission

The total weight of the unit is greater than in direct-drive or geared transmission. Turbine and diesel-electric units are around 20% heavier than comparable gearbox units. This extra 20% takes account of the fact that the primary energy producers must be larger to compensate for the losses in the electrical unit.

Development trends in engine plant weights

Engine weight has decreased as a result of higher supercharging. The weight of the electrical plant has increased corresponding to the increased electrical consumption. Engine room installations have increased due to automation, engine room insulation and heavy oil systems. (Heavy oil systems are considered worthwhile if the output exceeds 1000 kW and the operational time 3000 hours per year. The limit fluctuates according to the price situation.)

Centre of mass of the engine plant

The centre of mass of the engine plant is best determined using the weights of the individual groups.

The centre of mass of the main engine in trunk piston engines is situated at 0.35–0.45 of the height above the crank-shaft. In crosshead engines, the centre of mass lies at 0.30–0.35 of this remaining height.

Where the engine plant is not arranged symmetrically in the engine room, it is advisable to check the transverse position of the centre of mass. If the eccentricity of the centre of mass results in heeling angles greater than 1–2°, the weight distribution should be balanced. This can be done with the aid of settling tanks, which are nearly always filled. Where eccentricity is less marked, balancing can be effected via smaller storage tanks.

5.4 Weight margin

A reserve or design margin is necessary in the weight calculation for the following reasons:

1. Weight tolerances in parts supplied by outside manufacturers, e.g. in the thickness of rolled plates and in equipment components.
2. Tolerances in the details of the design, e.g. for cement covering in the peak tanks.
3. Tolerances in the design calculations and results.

A recommended weight margin is 3% of the deadweight of a new cargo ship. If the shipbuilder has little experience in designing and constructing the required type of ship, margins in weight and stability should be increased. This is particularly the case if a passenger ship is being built for the first time. If, however, the design is a reconstruction or similar to an existing ship, the margin can be reduced considerably. Smaller marginal weights are one advantage of series production.

Weight margins should be adequate but not excessive. Margins should not be applied simultaneously to individual weights and collective calculations as it is more appropriate to work with one easily controllable weight margin for all purposes.

It is also advisable to create a margin of stability with the weight margin by placing the centre of mass of the margin weight at around $1.2\overline{KG}$ above the keel. The weight margin can be placed at the longitudinal centre of mass G.

New regulations and trends in design lead to increasing weights especially for passenger ships.

5.5 References

CARREYETTE, J. (1978). Preliminary ship cost estimation. *Trans. RINA*, p. 235
CARSTENS, H. (1967). Ein neues Verfahren zur Bestimmung des Stahlgewichts von Seeschiffen. *Hansa*, p. 1864
DUDSZUS, A. and DANCKWARDT, E. (1982). *Schiffstechnik*. Verlag Technik, p. 243
EHMSEN, E. (1963). Schiffswinden und Spille. *Handbuch der Werften*, Vol. **VII**. Hansa, p. 300
EHMSEN, E. (1970). Schiffsgetriebe, Kupplungen und Verstellpropeller. *Handbuch der Werften*, Vol. **X**. Hansa, p. 230
EHMSEN, E. (1974a). Schiffsantriebsdieselmotorenwinden und Spille. *Handbuch der Werften*, Vol. **XII**. Hansa, p. 220
EHMSEN, E. (1974b). Schiffsgetriebe. *Handbuch der Werften*, Vol. **XII**. Hansa, p. 250
FABARIUS, H. (1963). Leichtgut—Ladegeschirr. *Handbuch der Werften*. Vol. **VII**. Hansa, p. 168
HARVALD, S. A. and JENSEN, J. J. (1992). Steel weight estimation for ships. PRADS Conference, Newcastle. Elsevier Applied Science, p. 2.1047
HENSCHKE, W. (1952). *Schiffbautechnisches Handbuch*. 1st edn, Vol. **1**, Verlag Technik, p. 577
HENSCHKE, W. (1965). *Schiffbautechnisches Handbuch*. 2nd edn, Vol. **2**, Verlag Technik, pp. 465, 467
HOLLENBACH, U. (1994). Method for estimating the steel- and light ship weight in ship design. *ICCAS'94*, Bremen, p. 4.17
KERLEN, H. (1985). *Über den Einfluß der Völligkeit auf die Rumpfstahlkosten von Frachtschiffen.* IfS Rep. 456, Univ. Hamburg
KRAUSE, A. and DANCKWARDT, E. (1965). *Schiffbautechnisches Handbuch*. 2nd edn, Vol. **2**, Verlag Technik, pp. 97, 467
MILLER, D. (1968). *The Economics of Container Ship Subsystem*. Report No. 3, Univ. of Michigan
MÜLLER-KÖSTER, T. (1973). Ein Beitrag zur Ermittlung des Stahlgewichts von Aufbauten und Deckshäusern von Handelsschiffen im Entwurfsstadium. *Hansa*, p. 307

MURRAY, J. M. (1964–1965). Large bulk carriers. *Transactions of the Institution of Engineering and Shipbuilding Scotland, IESS*, **108**, p. 203

N. N. (1975). Winterschiffahrt und Eisbrecherhilfe in der nördlichen Ostsee. *Hansa*, p. 477

PROTZ, O. (1965), Diesel oder Turbine bei Großtankern, *Hansa*, p. 637

SATO, S. (1967). Effect of Principle Dimensions on Weight and Cost of Large Ships. Society of Naval Architects and Marine Engineers, New York Metropolitan Section

SCHNEEKLUTH, H. (1972). Zur Frage des Rumpfstahlgewichts und des Rumpfstahlschwerpunkts von Handelsschiffen. *Hansa*, p. 1554

SCHNEEKLUTH, H. (1985). *Entwerfen von Schiffen*. Koehler, p. 281

SCHREIBER, H. (1977). Statistische Untersuchungen zur Bemessung der Generatorleistung von Handelsschiffen. *Hansa*, p. 2117

WANGERIN, A. (1954). Elektrische Schiffsanlagen. *Handbuch der Werften*. Hansa, p. 297

WATSON, D. G. M. and GILFILLAN, A. W. (1977). Some ship design methods. *The Naval Architect*, **4**; *Transactions RINA*, **119**, p. 279

WEBERLING, E. (1963). Schiffsentwurf. *Handbuch der Werften*, Vol. **VII**, *Hansa*

WEBERLING, E. (1965). Schiffsentwurf. *Handbuch der Werften*, Vol. **VIII**, *Hansa*

6

Ship propulsion

We will limit ourselves here to ships equipped with propellers. Waterjets as alternative propulsive systems for fast ships, or ships operating on extremely shallow water are discussed by Merz (1993) and Kruppa (1994).

6.1 Interaction between ship and propeller

Any propulsion system interacts with the ship hull. The flow field is changed by the (usually upstream located) hull. The propulsion system changes, in turn, the flow field at the ship hull. These effects and the open-water efficiency of the propeller determine the propulsive efficiency η_D:

$$\eta_D = \eta_H \cdot \eta_0 \cdot \eta_R = \frac{R_T \cdot V_s}{P_D}$$

η_H = hull efficiency
η_0 = open-water propeller efficiency
η_R = relative rotative efficiency
P_D = delivered power at propeller
R_T = total calm-water resistance
V_s = ship speed

$\eta_D \approx 0.6$–0.7 for cargo ships
$\eta_D \approx 0.4$–0.6 for tugs

Danckwardt gives the following estimate (Henschke, 1965):

$$\eta_D = 0.836 - 0.000165 \cdot n \cdot \nabla^{1/6}$$

n is the propeller rpm and ∇ [m^3] the displacement volume. All ships checked were within $\pm 10\%$ of this estimate; half of the ships within $\pm 2.5\%$.
Keller (1973) gives:

$$\eta_D = 0.885 - 0.00012 \cdot n \cdot \sqrt{L_{pp}}$$

HSVA gave, for twin-screw ships in 1957:

$$\eta_D = 0.69 - 12\,000 \cdot \left(0.041 - \frac{V_s}{n \cdot D_P} \right)^3 \pm 0.02$$

Ship speed V_s in [kn], propeller diameter D_P in [m], $0.016 \leq V_s/(n \cdot D_P) \leq 0.04$.

The installed power P_B has to overcome in addition efficiency losses due to shafts and bearings:

$$P_B = \eta_S \cdot P_D$$

The shaft efficiency η_S is typically 0.98–0.985.

The hull efficiency η_H combines the influence of hull–propeller interaction:

$$\eta_H = \frac{1 - t}{1 - w}$$

Thrust deduction fraction t and wake fraction w are discussed in more detail below.

For small ships with rake of keel, Helm (1980) gives an empirical formula:

$$\eta_H = 0.895 - \frac{0.0065 \cdot L}{\nabla^{1/3}} - 0.005 \cdot \frac{B}{T} - 0.033 \cdot C_P + 0.2 \cdot C_M + 0.01 \cdot \text{lcb}$$

lcb is here the longitudinal centre of buoyancy taken from $L_{pp}/2$ in [%L_{pp}].

The basis for this formula covers $3.5 \leq L/\nabla^{1/3} \leq 5.5$, $0.53 \leq C_P \leq 0.71$, $2.25 \leq B/T \leq 4.50$, $0.60 \leq C_M \leq 0.89$, rake of keel 40%T, $D_P = 0.75T$. T is taken amidships.

Thrust deduction

The thrust T measured in a propulsion test is higher than the resistance R_T measured in a resistance test (without propeller). So the propeller induces an additional resistance:

1. The propeller increases the flow velocities in the aftbody of the ship which increases frictional resistance.
2. The propeller decreases the pressure in the aftbody, thus increasing the inviscid resistance.

The second mechanism dominates for usual propeller arrangements. The thrust deduction fraction t couples thrust and resistance:

$$t = \frac{T - R_T}{T} \quad \text{or} \quad T(1 - t) = R_T$$

t is usually assumed to be the same for model and ship, although the friction component introduces a certain scale effect. Empirical formulae for t are:

For single-screw ships:

$t = 0.5 \cdot C_P - 0.12$, Heckscher for cargo ships

$t = 0.77 \cdot C_P - 0.30$, Heckscher for trawlers

$t = 0.5 \cdot C_B - 0.15$, Danckwardt for cargo ships

$t = w \cdot (1.57 - 2.3 \cdot C_B/C_{WP} + 1.5 \cdot C_B)$, SSPA for cargo ships

$t = 0.001979 \cdot L/(B(1 - C_P)) + 1.0585 \cdot B/L - 0.00524 - 0.1418D^2/(BT)$,
   Holtrop and Mennen (1978)

For twin-screw ships:

$t = 0.5 \cdot C_P - 0.18$, Heckscher for cargo ships

$t = 0.52 \cdot C_B - 0.18$, Danckwardt for cargo ships

$t = w \cdot (1.67 - 2.3 \cdot C_B/C_{WP} + 1.5 \cdot C_B)$, SSPA for cargo ships

$t = 0.325 \cdot C_B - 0.1885 \cdot D_P/\sqrt{B \cdot T}$, Holtrop and Mennen (1978)

Alte and Baur (1986) give an empirical coupling between t and the wake fraction w:

$$(1 - t) = (1 - w)^{0.4-0.8}$$

In general, in the early design stage it cannot be determined which t will give the best hull efficiency η_H. t can be estimated only roughly in the design stage and all of the above formulae have a much larger uncertainty margin than those for w given below. t thus represents the largest uncertainty factor in the power prognosis.

Wake

The wake is usually decomposed into three components:

- Friction wake
  Due to viscosity, the flow velocity relative to the ship hull is slowed down in the boundary layer, leading, in regions of high curvature (especially in the aftbody) to flow separation.
- Potential wake
  In an ideal fluid without viscosity and free surface, the flow velocity at the stern resembles the flow velocity at the bow, featuring lower velocities with a stagnation point.
- Wave wake
  The steady wave system of the ship changes locally the flow as a result of the orbital velocity under the waves. A wave crest above the propeller increases the wake fraction, a wave trough decreases it.

For the usual single-screw ships, the frictional wake dominates. Wave wake is only significant for $F_n > 0.3$ (Alte and Baur, 1986).

The measured wake fraction in model tests is larger than in full scale as boundary layer and flow separation are relatively larger in model scale. Correction formulae try to consider this overprediction, but the influence of separation can only be estimated and this often introduces a significant error margin. The errors in predicting the required power remain nevertheless small, as the energy loss due to the wake is partially recovered by the propeller. However, the errors in predicting the wake propagate completely when computing optimal propeller rpm and pitch.

Model tests feature relatively thicker boundary layers and stronger separation than full-scale ships. Consequently the model wake is more pronounced than the full-scale wake. However, this hardly affects the power prognosis, as part of the greater energy losses in the model are regained by the propeller. Errors in correcting the wake for full scale affect mostly the rpm or pitch of the propeller. Proposals to modify the shape of the model to partially correct for the differences of model and full-scale boundary layers (Schneekluth, 1994) have not been implemented.

The propeller action accelerates the flow field, again by typically 5–20%. The wake distribution is either measured by laser-doppler velocimetry or computed by CFD (see Section 2.11). While CFD is not yet capable of reproducing the wake with sufficient accuracy, the integral of the wake over the propeller plane, the wake fraction w, is predicted well. In the early design stage, the following empirical formulae may help to estimate the wake fraction:

For single-screw ships, Schneekluth (1988) gives, for cargo ships with stern bulb:

$$w = 0.5 \cdot C_P \cdot \frac{1.6}{1 + D_P/T} \cdot \frac{16}{10 + L/B}$$

Other formulae for single-screw ships are:

$w = 0.75 \cdot C_B - 0.24$, Krüger (1976)

$w = 0.7 \cdot C_P - 0.18$, Heckscher for cargo ships

$w = 0.77 \cdot C_P - 0.28$, Heckscher for trawlers

$w = 0.25 + 2.5(C_B - 0.6)^2$, Troost for cargo ships

$w = 0.5 \cdot C_B$, Troost for coastal feeders

$w = C_B/3 + 0.01$, Caldwell for tugs with $0.47 \le C_B \le 0.56$

$w = 0.165 \cdot C_B \cdot (\nabla^{1/3}/D_P) - 0.1 \cdot (F_n - 0.2)$, Papmehl

$$w = \frac{3}{1 - (C_P/C_{WP})^2} \cdot \frac{B}{L} \cdot \frac{E}{T} \cdot \left[1 - \frac{1.5 \cdot D + (\varepsilon + r)}{B} \right], \text{Telfer for cargo ships}$$

ε is the skew angle in radians, r is the rake angle in radians, E is height of the shaft centre over keel.

For twin-screw ships:

$w = 0.81 \cdot C_B - 0.34$, Krüger (1976) for cargo ships

$w = 0.7 \cdot C_P - 0.3$, Heckscher for cargo ships

$w = C_B/3 - 0.03$, Caldwell for tugs with $0.47 \le C_B \le 0.56$

Holtrop and Mennen (1978) and Holtrop (1984) give further more complicated formulae for w for single-screw and twin-screw ships, which can be integrated in a power prognosis program.

All the above formulae consider only a few main parameters, but the shape of the ship, especially the aftbody, influences the wake considerably. Other important parameters are propeller diameter and propeller clearance, which are unfortunately usually not explicitly represented in the above formulae. For bulk carriers with $C_B \approx 0.85$, $w < 0.3$ have been obtained by form optimization. The above formulae can thus predict too high w values for full ships.

Relative rotative efficiency

Theoretically, the relative rotative efficiency η_R accounts for the differences between the open-water test and the inhomogeneous three-dimensional propeller inflow encountered in a propulsion test. In reality, the propeller efficiency behind the ship cannot be measured and all effects not included in the hull efficiency, i.e. wake and thrust deduction fraction, are included in η_R. In addition, the partial recovery of rotational energy by the rudder contributes to η_R. This mixture of effects makes it difficult to express η_R as a function of a few defined parameters.

Holtrop and Mennen (1978) and Holtrop (1984) give

$\eta_R = 0.9922 - 0.05908 \cdot A_E/A_0 + 0.07424 \cdot (C_P - 0.0225 \cdot \text{lcb})$ for single-screw ships

$\eta_R = 0.9737 + 0.111 \cdot (C_P - 0.0225 \cdot \text{lcb}) - 0.06325 \cdot P/D_P$ for twin-screw ships

lcb is here the longitudinal centre of buoyancy taken from $L_{wl}/2$ in $[\%L_{wl}]$
A_E/A_0 is the blade area ratio of the propeller
P/D_P is the pitch-to-diameter ratio of the propeller

Helm (1980) gives for small ships:

$$\eta_R = 0.826 + 0.01 \frac{L}{\nabla^{1/3}} + 0.02 \frac{B}{T} + 0.1 \cdot C_M$$

The basis is the same as for Helm's formula for η_H.

$\eta_R = 1 \pm 0.05$ for propeller propulsion systems; Alte and Baur (1986) recommend, as a simple estimate, $\eta_R = 1.00$ for single-screw ships, $\eta_R = 0.98$ for twin-screw ships.

Jensen (1994) gives $\eta_R = 1.02$–1.06 for single-screw ships depending also on details of the experimental and correlation procedure.

6.2 Power prognosis using the admiralty formula

The 'admiralty formula' is still used today, but only for a very rough estimate:

$$P_B = \frac{\Delta^{2/3} \cdot V^3}{C}$$

The admiralty constant C is assumed to be constant for similar ships with similar Froude numbers, i.e. ships that have almost the same C_B, C_P, ∇/L, F_n, ∇, etc. Typical values for C in $[t^{2/3} \cdot kn^3/kW]$ are:

general cargo ships	400–600
bulker and tanker	600–750
reefer	550–700
feedership	350–500
warship	150

These values give an order of magnitude only. The constant C should be determined individually for basis ships used in design. Völker (1974) gives a modified admiralty formula for cargo ships with smaller scatter for C:

$$P_D = \frac{\Delta^{0.567} \cdot V^{3.6}}{C \cdot \eta_D}$$

η_D in this formula may be estimated by one of the above-mentioned empirical formulae. Strictly speaking, the exponent of V should be a function of speed range and ship hull form. The admiralty formula is thus only useful if a ship of the same type, size and speed range is selected to determine C. It is possible to increase the accuracy of the Völker formula by adjusting it to specific ship types.

More accurate methods to estimate the power requirements estimate the resistance as described below:

$$P_B = \frac{R_T \cdot V}{\eta_D \cdot \eta_S}$$

MacPherson (1993) provides some background and guidance to designers for simple computer-based prediction methods, and these are recommended for further studies.

6.3 Ship resistance under trial conditions

Decomposition of resistance

As the resistance of a full-scale ship cannot be measured directly, our knowledge about the resistance of ships comes from model tests. The measured calm-water resistance is usually decomposed into various components, although all these components usually interact and most of them cannot be measured individually. The concept of resistance decomposition helps in designing the hull form as the designer can focus on how to influence individual resistance components. Larsson and Baba (1996) give a comprehensive overview of modern methods of resistance decomposition (Fig. 6.1).

The total calm-water resistance of a new ship hull can be decomposed as

$$R_T = R_F + R_W + R_{PV}$$

It is customary to express the resistance by a non-dimensional coefficient, e.g.

$$C_T = \frac{R_T}{\rho/2 \cdot V^2 \cdot S}$$

S is the wetted surface, usually taken at calm-water conditions, although this is problematic for fast ships.

Empirical formulae to estimate S are:

For cargo ships and ferries (Lap, 1954):

$$S = \nabla^{1/3} \cdot (3.4 \cdot \nabla^{1/3} + 0.5 \cdot L_{WL})$$

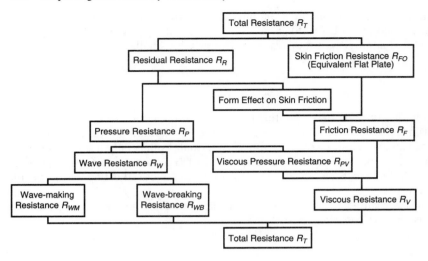

Figure 6.1 Decomposition of ship resistance components

For cargo ships and ferries (Danckwardt, 1969):

$$S = \frac{\nabla}{B} \cdot \left[\frac{1.7}{C_B - 0.2 \cdot (C_B - 0.65)} + \frac{B}{T} \right]$$

For trawlers (Danckwardt, 1969):

$$S = \frac{\nabla}{B} \cdot \left[\frac{1.7}{C_B} + \frac{B}{T} \cdot \left(0.92 + \frac{0.092}{C_B} \right) \right]$$

For modern warships (Schneekluth, 1988):

$$S = L \cdot (1.8 \cdot T + C_B \cdot B)$$

Friction resistance

The friction resistance is usually estimated taking the resistance of an 'equivalent' flat plate of the same area and length as reference:

$$R_F = C_F \cdot \frac{\rho}{2} \cdot V^2 \cdot S$$

$C_F = 0.075/(\log R_n - 2)^2$ according to ITTC 1957. The ITTC formula for C_F includes not only the flat plate friction, but also some form and roughness effects. C_F is a function of speed, shiplength, temperature and viscosity of the water. However, the speed dependence is almost negligible. For low speeds, friction resistance dominates. The designer will then try to keep the wetted surface S small. This results in rather low L/B and L/T ratios for bulkers and tankers.

Viscous pressure resistance

A deeply submerged model of a ship will have no wave resistance. But its resistance will be higher than just the frictional resistance. The form of the ship induces a local flow field with velocities that are sometimes higher and sometimes lower than the average velocity. The average of the resulting shear stresses is then higher. Also, energy losses in the boundary layer, vortices and flow separation prevent an increase to stagnation pressure in the aftbody as predicted in an ideal fluid theory. The viscous pressure resistance increases with fullness of waterplane and block coefficient.

An empirical formula for the viscous pressure resistance coefficient is (Schneekluth, 1988):

$$C_{PV} \cdot 10^3 = (26 \cdot C_\nabla + 0.16) + \left(\frac{B}{T} - \frac{13 - 10^3 \cdot C_\nabla}{6} \right)$$
$$\cdot (C_P + 58 \cdot C_\nabla - 0.408) \cdot (0.535 - 35 \cdot C_\nabla)$$

where $C_\nabla = \nabla/L^3$. The formula was derived from the Taylor experiments based on $B/T = 2.25$–4.5, $C_P = 0.48$–0.8, $C_\nabla = 0.001$–0.007.

This viscous pressure resistance is often written as a function of the friction resistance:

$$R_{PV} = k \cdot R_F$$

This so-called form factor approach does not properly include the separation effects. For slender ships, e.g. containerships, the resistance due to separation is negligible (Jensen, 1994). For some icebreakers, inland vessels and other ships with very blunt bows, the form factor approach appears to be inappropriate.

There are various formulae to estimate k:

$k = 18.7 \cdot (C_B \cdot B/L)^2$	Granville (1956)
$k = 14 \cdot (\nabla/L^3) \cdot (B/T)$	Russian, in Alte and Baur (1986)
$k = -0.095 + 25.6 \cdot C_B/[(L/B)^2 \cdot \sqrt{B/T}]$	Watanabe

The viscous pressure resistance depends on the local shape and CFD can be used to improve this resistance component.

Wave resistance

The ship creates a typical wave system which contributes to the total resistance. For fast, slender ships this component dominates. In addition, there are breaking waves at the bow which dominate for slow, full hulls, but may also be considerable for fast ships. The interaction of various wave systems is complicated leading to non-monotonous function of the wave resistance coefficient C_W. The wave resistance depends strongly on the local shape. Very general guidelines (see Sections 2.2 to 2.4, 2.9) and CFD (see Section 2.11) are used to improve wave resistance. Slight form changes may result in considerable improvements. This explains the margins of uncertainties for simple predictions of ship total resistance based on a few parameters as described below.

Prediction methods

Design engineers need simple and reasonably accurate estimates of the power requirements of a ship. Such methods focus on the prediction of the resistance. Some of the older methods listed below are still in use

- 'Ayre' for cargo ships, Remmers and Kempf (1949)
- 'Taggart' for tugboats
- 'Series-60' for cargo ships, Todd *et al.* (1957)
- 'BSRA' for cargo ships, Moor *et al.* (1961)
- 'Danckwardt' for cargo ships and trawlers, Danckwardt (1969)
- 'Helm' for small ships, Helm (1964)
- 'Lap–Keller' for cargo ships and ferries, Lap (1954), Keller (1973)

The following methods have general applicability:

- 'Taylor–Gertler' (for slender ships), Gertler (1954)
- 'Guldhammer–Harvald', Guldhammer and Harvald (1974)
- 'Holtrop–Mennen', Holtrop and Mennen (1978, 1982), Holtrop (1977, 1978, 1984)
- 'SSPA', Williams (1969)
- 'Hollenbach', Hollenbach (1997, 1998)

The older methods usually do not consider a bulbous bow. The effect of a bulbous bow may then be approximately introduced by increasing the length in the calculation by 2/3 of the bulb length.

   Tables 6.1 to 6.8 show an overview of some of the older methods. The resistance of modern ships is usually higher than predicted by the above methods. The reason is that the following modern form details increase resistance:

- Stern bulb.
- Hollow waterlines in the vicinity of the upper propeller blades to reduce thrust deduction.
- Large propeller aperture to reduce propeller induced vibrations.
- Immersed transom stern.
- Very broad stern to accommodate a stern ramp in ro-ro ships or to increase stability.
- V sections in the forebody of containerships to increase deck area.
- Compromises in the location of the shoulders to increase container stowage capacity.

The first two items improve propulsive efficiency. Thus power requirements may be lower despite higher resistance.

   The next section describes briefly Hollenbach's method, as this is the most modern, easily programmed and at least as good as the above for modern hull forms.

Table 6.1 Resistance procedure 'Ayre'

Year published: 1927, 1948

Basis for procedure: Evaluation of test results and trials

Description of main value

$C = \Delta^{0.64} \cdot V^3/P_E$ as $f(F_n, L/\Delta^{1/3})$

Target value

Effective power P_E [HP]

Input values

L_{pp}, $F_n = V/\sqrt{g \cdot L_{pp}}$; $\Delta^{0.64}$, $L/\Delta^{1/3}$; $C_{B,pp}$, B/T; lcb, L_{wl}

Range of variation of input values

$L_{pp} > 30\,\text{m}$; $0.1 \leq F_n \leq 0.3$; $0.53 \leq C_{B,PP} \leq 0.85$; $-2.5\%L \leq \text{lcb} \leq 2\%L$

Remarks

1. Influence of bulb not taken into account.
2. Included in the procedure are:
   (a) friction resistance using Froude
   (b) 8% additions for wind and appendages
   (c) relating to trial conditions.
3. The procedure is not applicable for $F_n > 0.3$ and usually yields higher values than other calculation methods.
4. Area of application: Cargo ships.
5. Constant or dependent variable values: $\beta = f(F_n)$.

References

WENDEL, K. (1954). Angenäherte Bestimmung der notwendigen Maschinenleistung. *Handbuch der Werften*, p. 34

Table 6.2 Resistance procedure 'Taylor-Gertler'

Year published: 1910, 1954, 1964

Basis for procedure: Systematic model tests with a model warship (Royal Navy armoured cruiser *Leviathan*)

Description of main value

1. *Gertler:* $C_R = R_R/((\rho/2) \cdot V^2 \cdot S)$ as $f(B/T, C_P, T_q$ or $F_n, \nabla/L_{wl}^3)$
2. *Rostock:* C_R as $f(C_P, \nabla/L_{wl}^3, F_n)$ for $B/T = 4.5$ and R_R/Δ [kp/Mp] as $f(B/T, \nabla/L_{wl}^3, F_n, C_P)$

Target value: Residual resistance R_R [kp]

Input values

$$L_{wl}; \ F_{n,WL} = \frac{V}{\sqrt{g \cdot L_{wl}}}; \ C_{P,WL}; \ \nabla/L_{wl}^3; \ B/T; \ S$$

Range of variation of input values

1. *Gertler:*
   $0.15 \leq F_n \leq 0.58; \ 2.25 \leq B/T \leq 3.75; \ 0.48 \leq C_P \leq 0.86; \ 0.001 \leq \nabla/L_{wl}^3 \leq 0.007$
2. *Rostock:*
   $0.15 \leq F_n \leq 0.33; \ 2.25 \leq B/T \leq 3.75; \ 0.48 \leq C_P \leq 0.86; \ 0.002 \leq \nabla/L_{wl}^3 \leq 0.007$

Remarks

1. Influence of bulb not taken into account.
2. The procedure generally underestimates by 5–10%.
3. Area of application: fast cargo ships, warships.
4. Constant or dependent variable values: $C_M = 0.925 =$ constant, $C_m =$ constant, lcb $= 0.5L_{wl}$.

References

GERTLER, M. (1954). A reanalysis of the original test data for the Taylor standard series. DTMB report 806, Washington

KRAPPINGER, O. (1963). Schiffswiderstand und Propulsion. *Handbuch der Werften*, Vol. **VII**, p. 118

HENSCHKE, W. (1957). *Schiffbautechnisches Handbuch* Vol. **1**, p. 353

HÄHNEL, G. and LABES, K. H. (1964). Systematische Widerstandsversuche mit Taylor-Modellen mit einem Breiten-Tiefgangsverhältnis B/T = 4.50. *Schiffbauforschung*, p. 123

Table 6.3 Resistance procedure 'Lap–Keller'

Year published: 1954, 1973

Basis for procedure: Evaluation of resistance tests (non-systematic) conducted at MARIN (Netherlands)

Description of main value: Resistance coefficient $C_R = R_R/((\rho/2) \cdot V^2 \cdot S$ as f(Group [lcb; C_P], Number of screws and B/T, $V/\sqrt{C_P \cdot L_{pp}}$)

Target value: Residual resistance R_R

Input values: L_{pp}; lcb; C_P; $V/\sqrt{C_P \cdot L_{pp}}$; Number of screws; B/T; $A_M = C_M \cdot B \cdot T \cdot S$

Range of variation of input values

$0.4 \leq V/\sqrt{C_P \cdot L_{pp}} \leq 1.5$; $0.55 \leq C_P \leq 0.85$; $-4\% \leq \text{lcb}/L_{pp} \leq 2\%$

Remarks

1. Influence of bulb not taken into account.
2. The procedure is highly reliable for the region specified.
3. Area of application: cargo and passenger ships

References

LAP, A. J. W. (1954). Diagrams for determining the resistance of single-screw ships. *International Shipbuilding Progress*, p. 179
HENSCHKE, W. (1957). *Schiffbautechnisches Handbuch* Vol. **2**, p. 129, p. 279
KELLER, W. H. auf'm (1973). Extended diagrams for determining the resistance and required power for single-screw ships. *International Shipbuilding Progress*, p. 133

Table 6.4 Resistance procedure 'Danckwardt'

Year published: 1969

Basis for procedure: Evaluation of model test series and individual tests

Description of main value: Specific resistance R_T/Δ as $f(L/B), B/T, F_n, C_B)$ for cargo and passenger ships, as $f(L/B), B/T, F_n, C_P)$ for stern trawlers

Target value: Total resistance R_T

Input values

L_{pp}; L_{pp}/B; B/T; $F_n = V/\sqrt{g \cdot L_{pp}}$; C_B for cargo and passenger ships; C_P for stern trawlers; C_A (roughness); (temperature of seawater and fresh water); lcb; frame form in fore part of ship; A_{BT} (section area at forward perpendicular); $S \cdot L_{pp}/\nabla$

Range of variation of input values

Cargo and pass. ships: $6 \leq L/B \leq 8$; $0.14 \leq F_n \leq 0.32$; $2 \leq B/T \leq 3$; $0.525 \leq C_B \leq 0.825$; $50\,\text{m} \leq L_{pp} \leq 280\,\text{m}$; $5°C \leq t° \leq 30°C$; $0.01 \leq A_{BT}/A_M \leq 0.15$

Stern trawlers

$4 \leq L/B \leq 7$; $0.1 \leq F_n \leq 0.36$; $2 \leq B/T \leq 3$; $0.55 \leq C_P \leq 0.7$; $25\,\text{m} \leq L_{pp} \leq 100\,\text{m}$; $-0.05L_{pp} \leq \text{lcb} \leq 0$

Remarks

1. Influence of bulb taken into account.
2. The procedure is highly reliable for the region specified.
3. Area of application: cargo and passenger ships, stern trawlers.
4. The Δ in the expression R_T/Δ is a weight 'force' of the ship, i.e. displacement mass times gravity acceleration.

References

DANCKWARDT, E. C. M. (1969). Ermittlung des Widerstands von Frachtschiffen und Hecktrawlern beim Entwurf. *Schiffbauforschung*, p. 124, Errata p. 288
DANCKWARDT, E. C. M. (1981). Algorithmus zur Ermittlung des Widerstands von Hecktrawlern. *Seewirtschaft*, p. 551
DANCKWARDT, E. C. M. (1985). Algorithmus zur Ermittlung des Widerstands von Frachtschiffen. *Seewirtschaft*, p. 390
DANCKWARDT, E. C. M. (1985). Weiterentwickeltes Verfahren zur Vorausberechnung des Widerstandes von Frachtschiffen. *Seewirtschaft*, p. 136

Table 6.5 Resistance procedure 'Series-60', Washington

Year published: 1951–1960

Basis for procedure: Systematic model tests with variations of five basic forms. Each basic form represents a block coefficient between 0.6 and 0.8. The basic form for $C_B = 0.8$ was specially designed for this purpose. The other basic forms were based on existing ships.

Description of main value

$$ © = \frac{427 \cdot P_E}{\Delta^{2/3} \cdot V_{kn}^3} \text{ as } f(B/T, L/B, ®, C_{B,pp}) $$

Target value: Total resistance R_t [kp]

Input values

$$ L_{pp}; C_{B,pp}; L_{pp}/B; B/T; ® = \sqrt{4\pi} \cdot V/\sqrt{g\nabla^{1/3}}; F_n = V/\sqrt{g \cdot L_{pp}} $$

Range of variation of input values

$5.5 \leq L/B \leq 8.5; 0.6 \leq C_{B,pp} \leq 0.8; 2.5 \leq B/T \leq 3.5; 1.2 \leq ® \leq 2.4; 45\,\text{m} \leq L_{pp} \leq 330\,\text{m}$

Remarks

1. Influence of bulb not taken into account.
2. In addition to resistance, propulsion, partial loading, trim and stern form were also investigated. This is the main advantage of this procedure.
3. Area of application: Cargo ships, tankers
4. Dependent values; constant or variable lcb= $f(C_{B,pp})$ and $C_M = f(C_{B,pp})$
5. The investigated forms differ considerably from modern hull forms.

The ship forms do not represent modern ship hulls. The greatest value of these series from today's view lies in the investigation of partial loading, trim and propulsion.

References

Transactions of the Society of Naval Architects and Marine Engineers 1951, 1953, 1954, 1956, 1957, 1960
Handbuch der Werften Vol. **VII**, p. 120
HENSCHKE, W. (1957), *Schiffbautechnisches Handbuch* Vol. **2**, pp. 135, 287
SABIT, A. S. (1972). An analysis of the Series 60 results, Part 1, Analysis of form and resistance results. *International Shipbuilding Progress*, p. 81
SABIT, A. S. (1972). An analysis of the Series 60 results, Part 2, Regression analysis of the propulsion factors. *International Shipbuilding Progress*, p. 294

Table 6.6 Resistance procedure 'SSPA', Gothenborg

Year published: 1948–1959 (summarized 1969)

Basis for procedure: Systematic model tests with ships of selected block coefficients

Description of main value

1. Residual resistance coefficient: $C_R = R_R/((\rho/2) \cdot V^2 \cdot S)$ as $f(C_{B,pp}, L/\nabla^{1/3}, F_n)$.
2. Friction resistance coefficient: $C_F = R_F/((\rho/2) \cdot V^2 \cdot S)$ as V_{kn}, L_{pp}.
3. Effective power [HP] as $f(V, C_{B,pp}, \nabla, L/\nabla^{1/3}, L_{pp})$.

Target value: Total resistance R_t [kp]

Input values

L_{pp}; $C_{B,pp}$; ∇; $F_n = V/\sqrt{g \cdot L_{pp}}$; $L/\nabla^{1/3}$

Range of variation of input values

$0.525 \leq C_{B,pp} \leq 0.75$; $1.5 \leq B/T \leq 6.5$; $80\,\text{m} \leq L_{pp} \leq 220\,\text{m}$ $0.18 \leq F_n \leq 0.32$; $5 \leq L/\nabla^{1/3} \leq 7$

Remarks

1. Influence of bulb not taken into account.
2. The second reference gives propulsion results.
3. Area of application: cargo ships, passenger ships.
4. Dependent values; constant or variable lcb= $f(C_{B,pp})$ and $C_M = f(C_{B,pp})$

References

Information from the Gothenborg research institute No. 66 (by A. Williams)
Information from the Gothenborg research institute No. 67

Table 6.7 Resistance procedure 'Taggart'

Year published: 1954

Basis for procedure: Systematic model tests

Description of main value: Residual resistance coefficient $C_R = R_R/((\rho/2) \cdot V^2 \cdot S)$ as $f(C_P, F_n, \nabla/L^3)$

Target value: Residual resistance R_R [kp]

Input values

L_{pp}, $F_n = V/\sqrt{g \cdot L_{pp}}$, C_P, ∇/L_{pp}^3

Range of variation of input values

$0.18 \leq F_n \leq 0.42$; $0.56 \leq C_P \leq 0.68$; $0.007 \leq \nabla/L_{pp}^3 \leq 0.015$

Remarks

1. The graph represents the continuation of the Taylor tests for $\nabla/L_{pp}^3 \geq 0.007$, but related to L_{pp}.
2. Area of application: tugs, fishing vessels.

References

Transactions of the Society of Naval Architects and Marine Engineers 1954, p. 632
HENSCHKE, W. (1957), *Schiffbautechnisches Handbuch* Vol. **2**, p. 1000

Table 6.8 Resistance procedure 'Guldhammer–Harvald'

Year published: 1965, 1974

Basis for procedure: Evaluation of well-known resistance calculation procedures (Taylor, Lap, Series 60, Gothenborg, BSRA, etc.)

Description of main value: Residual resistance coefficient $C_R = R_R/((\rho/2) \cdot V^2 \cdot S)$ as $f(F_{nWL}$ or $V/\sqrt{L_{wl}}, L_{wl}/\nabla^{1/3}, C_{P,WL})$
Friction resistance coefficient $C_F = R_F/((\rho/2) \cdot V^2 \cdot S)$ as $f(L_{wl}, V_{kn})$

Target value: Total resistance R_T [kp]

Input values

$L_{wl}, F_{nWL} = V/\sqrt{g \cdot L_{wl}}, B/T$, lcb, frame form, A_{BT} (bulb), S, $C_{P,WL}$, $L_{wl}/\nabla^{1/3}$

Range of variation of input values

$0.15 \le F_{n,WL} \le 0.44$; $0.5 \le C_{P,WL} \le 0.8$; $4.0 \le L_{wl}/\nabla^{1/3} \le 8.0$; lcb before lcb standard; Correction for A_{BT} only for $0.5 \le C_{P,WL} \le 0.6$

Remarks

1. Influence of bulb taken into account.
2. Reference to length in WL.
3. Area of application: universal, tankers.
4. The correction for the centre of buoyancy appears (from area to area) overestimated.
5. The procedure underestimates resistance for ships with small L/B.

References

GULDHAMMER, H. E. and HARVALD, S. A. (1974). *Ship Resistance, Effect of Form and Principal Dimensions*. Akademisk Forlag, Copenhagen
HARVALD, S. A. (1978). Estimation of power of ships. *International Shipbuilding Progress*, p. 65
HENSCHKE, W. (1957). *Schiffbautechnisches Handbuch* Vol. **2**, p. 1000

Hollenbach's method

Hollenbach (1997, 1998) analysed model tank tests for 433 ships performed by the Vienna Ship Model Basin during the period from 1980 to 1995 to improve the reliability of the performance prognosis of modern cargo ships in the preliminary design stage. Hollenbach gives formulae for the 'best-fit' curve, but also a curve describing the lower envelope, i.e. the minimum a designer may hope to achieve after extensive optimization of the ship lines if its design is not subject to restrictions.

In addition to $L = L_{pp}$ and L_{wl}, which are defined as usual, Hollenbach uses a 'length over surface' L_{os} which is defined as follows:

- For design draft: length between aft end of design waterline and most forward point of ship below design waterline.
- For ballast draft: length between aft end and forward end of ballast waterline (rudder not taken into account).

Hollenbach gives the following empirical formulae to estimate the wetted surface including appendages:

$$S_{total} = k \cdot L \cdot (B + 2 \cdot T)$$

$$k = a_0 + a_1 \cdot L_{os}/L_{wl} + a_2 \cdot L_{wl}/L + a_3 \cdot C_B + a_4 \cdot B/T$$
$$+ a_6 \cdot L/T + a_7 \cdot (T_A - T_F)/L + a_8 \cdot D_P/T$$
$$+ k_{Rudd} \cdot N_{Rudd} + k_{Brac} \cdot N_{Brac} + k_{Boss} \cdot N_{Boss}$$

with coefficients according to Table 6.9.

Table 6.9 Coefficients for wetted surface in Hollenbach's method

	Single-screw		Twin-screw	
	design draft	ballast draft	bulbous bow	no bulbous bow
a_0	−0.6837	−0.8037	−0.4319	−0.0887
a_1	0.2771	0.2726	0.1685	0.0000
a_2	0.6542	0.7133	0.5637	0.5192
a_3	0.6422	0.6699	0.5891	0.5839
a_4	0.0075	0.0243	0.0033	−0.0130
a_5	0.0275	0.0265	0.0134	0.0050
a_6	−0.0045	−0.0061	−0.0006	−0.0007
a_7	−0.4798	0.2349	−2.7932	−0.9486
a_8	0.0376	0.0131	0.0072	0.0506
k_{Rudd}			0.0131	0.0076
k_{Brac}			−0.0030	−0.0036
k_{Boss}			0.0061	0.0049

D_P propeller diameter
T_A draft at aft perpendicular
T_F draft at forward perpendicular
N_{Rudd} number of rudders
N_{Brac} number of brackets
N_{Boss} number of bossings

Resistance

The resistance is decomposed without using a form factor.
The Froude number in the following formulae is based on the length L_{fn}:

$$L_{fn} = L_{os} \qquad\qquad\qquad L_{os}/L < 1$$
$$L_{fn} = L + 2/3 \cdot (L_{os} - L) \quad 1 \leq L_{os}/L < 1.1$$
$$L_{fn} = 1.0667 \cdot L \qquad\qquad 1.1 \leq L_{os}/L$$

The residual resistance is given by:

$$R_R = C_R \cdot \frac{\rho}{2} \cdot V^2 \cdot \left(\frac{B \cdot T}{10} \right)$$

Note that $(B \cdot T)/10$ is used instead of S as reference area. The non-dimensional coefficient C_R is generally expressed as:

$$C_R = C_{R,\text{Standard}} \cdot C_{R,\text{Fnkrit}} \cdot k_L \cdot (T/B)^{b1} \cdot (B/L)^{b2} \cdot (L_{os}/L_{wl})^{b3} \cdot (L_{wl}/L)^{b4}$$
$$\cdot (1 + (T_A - T_F)/L)^{b5} \cdot (D_P/T_A)^{b6} \cdot (1 + N_{\text{Rudd}})^{b7}$$
$$\cdot (1 + N_{\text{Brac}})^{b8} \cdot (1 + N_{\text{Boss}})^{b9} \cdot (1 + N_{\text{Thruster}})^{b10}$$

where N_{Thruster} is the number of side thrusters.

$$C_{R,\text{Standard}} = c_{11} + c_{12}F_n + c_{13}F_n^2 + C_B \cdot (c_{21} + c_{22}F_n + c_{23}F_n^2)$$
$$+ C_B^2 \cdot (c_{31} + c_{32}F_n + c_{33}F_n^2)$$
$$C_{R,\text{Fnkrit}} = \max(1.0, (F_n/F_{n,\text{krit}})^{f1})$$
$$F_{n,\text{krit}} = d_1 + d_2 C_B + d_3 C_B^2$$
$$k_L = e_1 L^{e2}$$

Typical resistance

The typical residual resistance coefficient is then determined by the coefficients in Table 6.10. The range of validity is given by Table 6.11. Table 6.12 gives the range of the standard mean deviation of the database considered. Within this range, the formulae should be reasonably accurate, but values outside this range may also be used.

Minimum resistance

Very good hulls, not subject to special design constraints enforcing hydrodynamically suboptimal hull forms, may achieve the following residual resistance coefficients:

$$C_R = C_{R,\text{Standard}} \cdot (T/B)^{a1} \cdot (B/L)^{a2} \cdot (L_{os}/L_{wl})^{a3} \cdot (L_{wl}/L)^{a4}$$

Table 6.13 gives the appropriate coefficients, Table 6.14 the range of validity.

Table 6.10 Coefficients for typical resistance in Hollenbach's method

	Single-screw		Twin-screw
	design draft	ballast draft	
$b1$	−0.3382	−0.7139	−0.2748
$b2$	0.8086	0.2558	0.5747
$b3$	−6.0258	−1.1606	−6.7610
$b4$	−3.5632	0.4534	−4.3834
$b5$	9.4405	11.222	8.8158
$b6$	0.0146	0.4524	−0.1418
$b7$	0	0	−0.1258
$b8$	0	0	0.0481
$b9$	0	0	0.1699
$b10$	0	0	0.0728
c_{11}	−0.57420	−1.50162	−5.34750
c_{12}	13.3893	12.9678	55.6532
c_{13}	90.5960	−36.7985	−114.905
c_{21}	4.6614	5.55536	19.2714
c_{22}	−39.721	−45.8815	−192.388
c_{23}	−351.483	121.820	388.333
c_{31}	−1.14215	−4.33571	−14.3571
c_{32}	−12.3296	36.0782	142.738
c_{33}	459.254	−85.3741	−254.762
d_1	0.854	0.032	0.897
d_2	−1.228	0.803	−1.457
d_3	0.497	−0.739	0.767
e_1	2.1701	1.9994	1.8319
e_2	−0.1602	−0.1446	−0.1237
f_1	$F_n/F_{n,\text{krit}}$	$10 \cdot C_B \cdot (F_n/F_{n,\text{krit}} - 1)$	$F_n/F_{n,\text{krit}}$

Table 6.11 Range of validity for typical resistance, Hollenbach's method

	Single-screw		Twin-screw
	design draft	ballast draft	
$F_{n,\min}$, $C_B \leq 0.6$	0.17	$0.15 + 0.1 \cdot (0.5 - C_B)$	0.16
$F_{n,\min}$, $C_B > 0.6$	$0.17 + 0.2 \cdot (0.6 - C_B)$	$0.15 + 0.1 \cdot (0.5 - C_B)$	$0.16 + 0.24 \cdot (0.6 - C_B)$
$F_{n,\max}$	$0.642 - 0.635 \cdot C_B + 0.15 \cdot C_B^2$	$0.32 + 0.2 \cdot (0.5 - C_B)$	$0.50 + 0.66 \cdot (0.5 - C_B)$

Table 6.12 Standard deviation of database for typical resistance, Hollenbach's method

	Single-screw		Twin-screw
	design draft	ballast draft	
$L/\nabla^{1/3}$	4.490–6.008	5.450–7.047	4.405–7.265
C_B	0.601–0.830	0.559–0.790	0.512–0.775
L/B	4.710–7.106	4.949–6.623	3.960–7.130
B/T	1.989–4.002	2.967–6.120	2.308–6.110
L_{os}/L_{wl}	1.000–1.050	1.000–1.050	1.000–1.050
L_{wl}/L	1.000–1.055	0.945–1.000	1.000–1.070
D_P/T_A	0.430–0.840	0.655–1.050	0.495–0.860

Table 6.13 Coefficients for minimum resistance in Hollenbach's method

	a1	a2	a3	a4
Single-screw ship	−0.3382	0.8086	−6.0258	−3.5632
Twin-screw ship	−0.2748	0.5747	−6.7610	−4.3834

For single-screw ships

a_{00}	−0.9142367	a_{10}	4.6614022	a_{20}	−1.1421462
a_{01}	13.389283	a_{11}	−39.720987	a_{21}	−12.329636
a_{02}	90.596041	a_{12}	−351.48305	a_{22}	459.25433

For twin-screw ships

a_{00}	3.2727938	a_{10}	−11.501201	a_{20}	12.462569
a_{01}	−44.113819	a_{11}	166.55892	a_{21}	−179.50549
a_{02}	171.69229	a_{12}	−644.45600	a_{22}	680.92069

Table 6.14 Range of validity for minimum resistance, Hollenbach's method

	Single-screw	Twin-screw
$F_{n,\min}$, $C_B \le 0.6$	0.17	0.15
$F_{n,\min}$, $C_B > 0.6$	$0.17 + 0.2 \cdot (0.6 - C_B)$	0.14
$F_{n,\max}$	$0.614 - 0.717 \cdot C_B + 0.261 \cdot C_B^2$	$0.952 - 1.406 \cdot C_B + 0.643 \cdot C_B^2$

6.4 Additional resistance under service conditions

Appendages

Properly arranged bilge keels contribute only 1–2% to the total resistance of ships. However, trim and ship motions in seastates increase the resistance more than for ships without bilge keels. Thus, in evaluation of model tests, a much higher increase of resistance should be made for ships in ballast condition.

Bow-thrusters, if properly designed and located, do not significantly increase resistance. Transverse thrusters in the aftbody may increase resistance by 1–6% (Brix, 1986).

Shaft brackets and bossings increase resistance by 5–12% (Alte and Baur, 1986). For twin-screw ships with long propeller shafts, the resistance increase maybe more than 20% (Jensen, 1994).

Rudders increase resistance little (∼1%) if in neutral position and improve propulsion. But even moderate rudder angles increase resistance by 2–6% (Alte and Baur, 1986).

Shallow water

Shallow water increases friction resistance and usually also wave resistance. Near the critical depth Froude number $F_{nh} = V/\sqrt{gh} = 1$, where h is the water depth, the resistance is strongly increased. Figure 6.2 allows one to estimate the speed loss for weak shallow-water influence (Lackenby, 1963). For strong shallow-water influence a simple correction is impossible as wave breaking, squat and deformation of the free surface introduce complex physical interactions. In this case, only model tests or to some extent CFD may help.

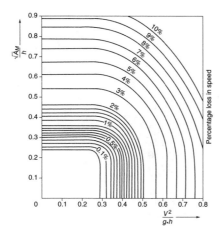

Figure 6.2 Shallow water influence and speed loss for shallow water

Wind

Wind resistance is important for ships with large lateral areas above the water level, e.g. containerships and car ferries. Fast and unconventional ships, e.g. air-cushioned vehicles, also require inclusion of the contribution of wind or air resistance. Jensen (1994) gives a very simple estimate for the wind resistance of cargo ships:

$$R_{AA} = C_{AA} \frac{\rho_{air}}{2} \cdot (V + V_{wind})^2 \cdot A_F$$

For cargo ships $C_{AA} = 0.8–1.0$. $\rho_{air} = 1.25 \, kg/m^3$ the density of air, V_{wind} is the absolute value of wind speed and A_F is the frontal projected area of the ship above sea level.

The wind resistance may be estimated with more accuracy following Blendermann (1993, 1996):

$$R_{AA} = \frac{\rho_{air}}{2} u^2 \cdot A_L \cdot CD_l \frac{\cos \varepsilon}{1 - \frac{\delta}{2}\left(1 - \frac{CD_l}{CD_t}\right) \sin^2 2\varepsilon}$$

where u is the apparent wind velocity, A_L the lateral-plane area, ε the apparent wind angle ($\varepsilon = 0°$ in head wind), δ the cross-force parameter, and coefficients CD_t and CD_l the non-dimensional drag in beam wind and head wind, respectively. It is convenient to give the longitudinal drag with respect to the frontal projected area A_F:

$$CD_{l\,AF} = CD_l \frac{A_L}{A_F}$$

Table 6.15 gives typical values for CD_t, $CD_{l\,AF}$ and δ. The maximum wind resistance usually occurs for $0° < \varepsilon < 20°$. The above formulae and the values in the table are for uniform or nearly uniform flow, e.g. above the ocean. The wind speed at a height of 10 m above sea level is usually taken as reference

Table 6.15 Coefficients to estimate wind resistance, Blendermann (1996)

	CD_t	CD_{lAF}	δ
Car carrier	0.95	0.55	0.80
Cargo ship, container on deck, bridge aft	0.85	0.65/0.55	0.40
Containership, loaded	0.90	0.55	0.40
Destroyer	0.85	0.60	0.65
Diving support vessel	0.90	0.60	0.55
Drilling vessel	1.00	0.70–1.00	0.10
Ferry	0.90	0.45	0.80
Fishing vessel	0.95	0.70	0.40
LNG tanker	0.70	0.60	0.50
Offshore supply vessel	0.90	0.55	0.55
Passenger liner	0.90	0.40	0.80
Research vessel	0.85	0.55	0.60
Speed boat	0.90	0.55	0.60
Tanker, loaded	0.70	0.90	0.40
Tanker, in ballast	0.70	0.75	0.40
Tender	0.85	0.55	0.65

speed. Wind speed in Beaufort (Beaufort number BN) is converted to [m/s] by:

$$u_{10} = 0.836 \cdot BN^{1.5}$$

Blendermann (1993) gives further details on wind forces, especially for side forces, yaw and roll moments.

Roughness

The friction resistance can increase considerably for rough surfaces (Naess, 1983). For newbuilds, the effect of roughness is included in the ITTC line or the correlation constant. The values of the correlation constant differ considerably between different towing tanks depending on the extrapolation procedures employed and are subject to continuing debate among hydrodynamicists. In general, correlation allowances decrease with ship size and may become negative for very large ships. For guidance, Table 6.16 recommends values in conjunction with the ITTC 1957 friction coefficients (Keller, 1973). Of course, there is no negative 'roughness' in reality. Rather, the correlation allowance includes other effects which dominate the roughness correction for large ships.

Table 6.16 Correlation allowance with ITTC line

L_{wl} [m]	100	180	235	280	325	400
C_A	0.0004	0.0002	0.0001	0	−0.0001	−0.00025

A rough hull surface (without fouling) increases the frictional resistance by up to 5% (Jensen, 1994). Fouling can increase the resistance by much more. However, modern paints prevent fouling to a large extent and are also 'self-polishing', i.e. the paint will not become porous like older paints. More extensive discussions of the influence of roughness can be found in Berger

(1983), Collatz (1984), and Alte and Baur (1986). For ship hull design, the problem of roughness is not important.

Seaway

The added resistance of a ship in a seaway may be determined by computational methods which are predominantly based on strip methods (Söding and Bertram, 1998). However, such predictions for a certain region or route depend on the accuracy of seastate statistics. Ship size is generally more important than ship shape, although a low C_B is deemed to be advantageous. Bales *et al.* (1980) give seastate statistics that can be recommended for the North Atlantic.

Townsin and Kwon (1983) give simple approximate formulae to estimate the speed loss due to added resistance in wind and waves:

$$\Delta V = C_\mu \cdot C_{\text{ship}} \cdot V$$

C_μ is a factor considering the predominant direction of wind and waves, depending on the Beaufort number BN:

$$C_\mu = 1.0 \qquad\qquad\qquad \text{for} \quad \mu = 0°\text{--}30°$$

$$C_\mu = 1.7 - 0.03 \cdot (\text{BN} - 4)^2 \quad \text{for} \quad \mu = 30°\text{--}60°$$

$$C_\mu = 0.9 - 0.06 \cdot (\text{BN} - 6)^2 \quad \text{for} \quad \mu = 60°\text{--}150°$$

$$C_\mu = 0.4 - 0.03 \cdot (\text{BN} - 8)^2 \quad \text{for} \quad \mu = 150°\text{--}180°$$

C_{ship} is a factor considering the ship type:

$$C_{\text{ship}} = 0.5\text{BN} + \text{BN}^{6.5}/(2.7 \cdot \nabla^{2/3}) \quad \text{for tankers, laden}$$

$$C_{\text{ship}} = 0.7\text{BN} + \text{BN}^{6.5}/(2.7 \cdot \nabla^{2/3}) \quad \text{for tankers, ballast}$$

$$C_{\text{ship}} = 0.7\text{BN} + \text{BN}^{6.5}/(2.2 \cdot \nabla^{2/3}) \quad \text{for containerships}$$

∇ is the volume displacement in [m³]. Table 6.17 gives relations between Beaufort number, wind speeds and average wave heights.

Table 6.17a Wind strengths in Beaufort (Bft),
Henschke (1965)

Bft	Wind description	Wind speed [m/s]
0	No wind	0.0–0.2
1	Gentle current of air	0.3–1.5
2	Gentle breeze	1.6–3.3
3	Light breeze	3.4–5.4
4	Moderate breeze	5.5–7.9
5	Fresh breeze	8.0–10.7
6	Strong wind	10.8–13.8
7	Stiff wind	13.9–17.1
8	Violent wind	17.2–20.7
9	Storm	20.8–24.4
10	Violent storm	24.5–28.3
11	Hurricane-like storm	28.5–32.7
12	Hurricane	>32.7

Table 6.17b Sea strengths for North Sea coupled to wind strengths, Henschke (1965)

			Approximate average	
Sea scale	Bft	Sea description	Wave height [m]	Wavelength [m]
0	0	Smooth sea	—	—
1	1	Calm, rippling sea	0–0.5	0–10
2	2–3	Gentle sea	0.5–0.75	10–12.5
3	4	Light sea	0.75–1.25	12.5–22.5
4	5	Moderate sea	1.25–2.0	22.5–37.5
5	6	Rough sea	2.0–3.5	37.5–60.0
6	7	Very rough sea	3.5–6.0	60.0–105.0
7	8–9	High sea	>6.0	>105.0
8	10	Very high sea	up to 20	up to 600
9	11–12	Extremely heavy sea	up to 20	up to 600

6.5 References

ALTE, R. and BAUR, M. v. (1986). Propulsion. *Handbuch der Werften*, Vol. **XVIII**, Hansa, p. 132

BALES, S. L., LEE, W. T., VOELKER, J. M. and BRANDT, W. (1980). Standardized wave and wind environments for Nato operation areas. DTNSRDC Report SPD-0919-01

BERGER, G. (1983). *Untersuchung der Schiffsrauhigkeit*. Rep. 139, Forschungszentrum des Deutschen Schiffbaus, Hamburg

BLENDERMANN, W. (1993). Parameter identification of wind loads on ships. *Journal of Wind Engineering and Industrial Aerodynamics* **51**, p. 339

BLENDERMANN, W. 1996. Wind loading of ships—Collected data from wind tunnel tests in uniform flow. IfS-Rep. 574, Univ. Hamburg

BRIX, J. (1986). Strahlsteuer. *Handbuch der Werften*, Vol. **XVIII**, Hansa, p. 80

COLLATZ, G. (1984). Widerstanderhöhung durch Außenhautrauhigkeit. *IfS Kontaktstudium*, Univ. Hamburg

DANCKWARDT, E. C. M. (1969). Ermittlung des Widerstandes von Frachtschiffen und Hecktrawlern beim Entwurf. *Schiffbauforschung*, p. 124

GERTLER, M. (1954). A reanalysis of the original test data for the Taylor standard series. DTMB report 806

GRANVILLE, P. S. (1956). The viscous resistance of surface vessles and the skin friction of flat plates. *Transactions of the Society of Naval Architects and Marine Engineers*, p. 209

GULDHAMMER, H. E. and HARVALD, S. A. (1974). *Ship Resistance, Effect of Form and Principal Dimensions*. Akademisk Forlag Copenhagen

HELM, G. (1964). Systematische Widerstands-Untersuchungen von Kleinschiffen. *Hansa*, p. 2179

HELM, G. (1980). Systematische Propulsions-Untersuchungen von Kleinschiffen. Rep. 100, Forschungszentrum des Deutschen Schiffbaus, Hamburg

HENSCHKE, W. (1965). *Schiffbautechnisches Handbuch*. 2nd edn, Verlag Technik

HOLLENBACH, K. U. (1997). Beitrag zur Abschätzung von Widerstand und Propulsion von Ein- und Zweischraubenschiffen im Vorentwurf. IfS-Rep. 588, Univ. Hamburg

HOLLENBACH, K. U. (1998). Estimating resistance and propulsion for single-screw and twin-screw ships. *Ship Technology Research* 45/2

HOLTROP, J. (1977). A statistical analysis of performance test results. *International Shipbuilding Progress*, p. 23

HOLTROP, J. (1978). Statistical data for the extrapolation of model performance. *International Shipbuilding Progress*, p. 122

HOLTROP, J. (1984). A statistical re-analysis of resistance and propulsion data. *International Shipbuilding Progress*, p. 272

HOLTROP, J. and MENNEN, G. G. (1978). A statistical power prediction method. *International Shipbuilding Progress*, p. 253

HOLTROP, J. and MENNEN, G. G. (1982). An approximate power prediction method. *International Shipbuilding Progress*, p. 166

JENSEN, G. (1994). Moderne Schiffslinien. *Handbuch der Werften*, Vol. **XXII**, Hansa, p. 93

KELLER, W. H. auf'm (1973). Extended diagrams for determining the resistance and required power for single-screw ships. *International Shipbuilding Progress*, p. 253

KRÜGER, J. (1976). Widerstand und Propulsion. *Handbuch der Werften*, Vol. **XIII**, Hansa, p. 13

KRUPPA, C. (1994). Wasserstrahlantriebe. *Handbuch der Werften*, Vol. **XXII**, Hansa, p. 111

LACKENBY, H. (1963) The effect of shallow water on ship speed. *Shipbuilder and Marine Engine-builder*, Vol. 70, p. 446

LAP, A. J. W. (1954). Diagrams for determining the resistance of single-screw ships. *International Shipbuilding Progress*, p. 179

LARSSON, L. and BABA, E. (1996). Ship resistance and flow computations. *Advances in Marine Hydrodynamics*, M. Ohkusu (ed.), Comp. Mech. Publ.

MacPHERSON, D. M. (1993). Reliable performance prediction: Techniques using a personal computer. *Marine Technology* **30/4**, p. 243

MERZ, J. (1993). Ist der Wasserstrahlantrieb die wirtschaftliche Konsequenz?, *Hansa*, p. 52

MOOR, D. I., PARKER, M. N. and PATULLO, R. N. M. (1961). The BSRA methodical series—An overall presentation. *Transactions RINA*, p. 329

NAESS, E. (1983). Surface roughness and its influence on ship performance. *Jahrbuch Schiffbautechn. Gesellschaft*, Springer, p. 125

REMMERS, K. and KEMPF, E. M. (1949). Bestimmung der Schleppleistung von Schiffen nach Ayre. *Hansa*, p. 309

SCHNEEKLUTH, H. (1988). *Hydromechanik zum Schiffsentwurf.* Koehler

SCHNEEKLUTH, H. (1994). Model similitude in towing tests. *Schiffstechnik*, p. 44

SÖDING, H. and BERTRAM, V. (1998). Schiffe im Seegang. *Handbuch der Werften*, Vol. **XXIV**, Hansa

TODD, F. H., STUNTZ, G. R. and PIER, P. C. (1957). Series 60—The effect upon resistance and power of variation in ship proportions. *Transactions of the Society of Naval Architects and Marine Engineers*, p. 445

TOWNSIN, R. L. and KWON, Y. J. (1983). Approximate formulae for the speed loss due to added resistance in wind and waves. *Transactions RINA*, p. 199

VÖLKER, H. (1974) Entwerfen von Schiffen. *Handbuch der Werften* Vol. **XII**, Hansa, p. 17

WILLIAMS, A. (1969). The SSPA cargo liner series-resistance. SSPA Rep. 66

Appendix

A.1 Stability regulations

Historical perspective: Rahola's criterion

Rahola (1939) analysed statistically accidents caused by defects in stability and included the results in recommendations for 'safe stability'. These recommendations are based on the criterion of a degree of dynamic stability up to 40° angle of heel. The dynamic stability can be represented by the area below the stability moment curve, i.e. as the integral of the stability moment over the range of inclination (Fig. A.1). (This quantity equals the mechanical work done, or energy used, in heeling the ship.) If the righting arm h is considered instead of the stability moment M_{St}, the area below the righting arm curve represents the dynamical lever e. This distance e is identical with the increase in the vertical distance between form and mass centres of gravity in heeled positions (Fig. A.2). e can be found by numerically evaluating the righting arm curve.

Rahola's investigation resulted in the standard requirements:

righting lever for 20° heel:	$h_{20°} \geq 0.14\,\text{m}$
righting lever for 30° heel:	$h_{30°} \geq 0.20\,\text{m}$
heel angle of maximum righting lever:	$\phi_{\max} \geq 35°$
range of stability:	$\phi_0 \geq 60°$

Other righting levers are seen as equivalent if

$$e = \int_0^{40°} h \, \mathrm{d}\phi \geq 0.08\,\text{m}$$

for $\phi_{\max} \geq 40°$, where $\phi_{\max}$ is the upper limit of integration (Fig. A.3).

Rahola's criterion disregards important characteristics (e.g. seakeeping behaviour) and was derived for small cargo ships, especially coasters of a type which prevailed in the 1930s in the Baltic Sea. Nevertheless, Rahola's criterion became and still is widely popular with statutory bodies. The Germanischer

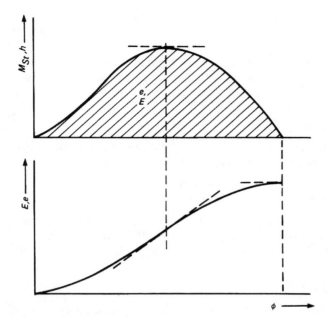

Figure A.1 Dynamic stability energy E

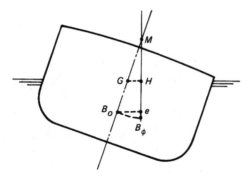

Figure A.2 Lever of dynamical stability $e = \overline{HB_\phi} - \overline{B_0G} \cos \phi$

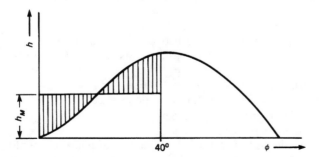

Figure A.3 Determining the dynamical lever e using Rahola

Lloyd confirmed the applicability of Rahola's criterion for standard post-war ships by analysing stability accidents which occurred after World War II (Seefisch, 1965).

While it was never made directly a stability regulation, Rahola's criterion has influenced most stability regulations for cargo ships and trawlers intended to guarantee a minimum safety against capsizing.

International regulations

Various stability requirements of the past have been consolidated into a few international codes on stability which apply for virtually all cargo ships:

- The Code on Intact Stability (IMO regulation A.749(18))
- SOLAS (1974) concerning damage stability

In addition, Rule 25 of MARPOL 73/78 affects damaged stability of tankers. This book reflects the state of the regulations in 1997. Modifications and additions are actively discussed. Stability regulations will thus undoubtedly change over time.

Code on Intact Stability

The Code on Intact Stability, IMO Resolution A.749(18), consolidates several previous stability regulations (IMO, 1995). The code contains regulations concerning all cargo ships exceeding 24 m in length with additional special rules for:

- cargo ships carrying timber deck cargo
- cargo ships carrying grain in bulk
- containerships
- passenger ships
- fishing vessels
- special purpose ships
- offshore supply vessels
- mobile offshore drilling units
- pontoons
- dynamically supported craft

The main design criteria of the code are:

- General intact stability criteria for all ships:
  1. $e_{0,30°} \geq 0.055$ m·rad; $e_{0,30°}$ is the area under the static stability curve to 30°
     $e_{0,40°} \geq 0.09$ m·rad; corresponding area up to 40°
     $e_{30,40°} \geq 0.03$ m·rad; corresponding area between 30° and 40°.
     If the angle of flooding ϕ_f is less than 40°, ϕ_f instead of 40° is to be used in the above rules.
  2. $h_{30°} \geq 0.20$ m; $h_{30°}$ is the righting lever at 30° heel.
  3. The maximum righting lever must be at an angle $\phi \geq 25°$.
  4. The initial metacentric height $\overline{GM}_0 \geq 0.15$ m.
- In addition, IMO requires for passenger ships:
  1. The heel angle on account of crowding of passengers to one side should not exceed 10°. A standard weight of 75 kg per passenger and four passengers/m² are assumed.

2. The heel angle on account of turning should not exceed $10°$. The heeling moment is

$$M_{Kr} = 0.02 \cdot \frac{V_0^2}{L} \cdot \Delta \cdot \left(\overline{KG} - \frac{T}{2} \right)$$

- Severe wind and rolling criterion (weather criterion):
  The weather criterion is intended to reflect the ability of the ship to withstand the combined effects of beam wind and rolling (Fig. A.4). The weather criterion requires that area $b \geq a$. The angles in Fig. A.4 are defined as follows:

  ϕ_0 angle of heel under action of steady wind; $16°$ or 80% of the angle of deck immersion, whichever is less, are suggested as maximum.

  ϕ_1 angle of roll windward due to wave action

  ϕ_2 minimum of ϕ_f, $50°$, ϕ_c

  ϕ_f is the heel angle at which openings in the hull, superstructures or deckhouses, which cannot be closed weathertight, immerse.

  ϕ_c angle of second intercept between wind heeling lever l_{w2} and righting arm curve.

  The wind heeling levers are constant at all heel angles:

$$l_{w1} = 0.051376 \, \frac{\text{kg}}{\text{m}^2} \frac{A \cdot Z}{\Delta}$$

$$l_{w2} = 1.5 \cdot l_{w1}$$

A is the projected lateral area of the portion of the ship and deck cargo above the waterline in $[\text{m}^2]$.

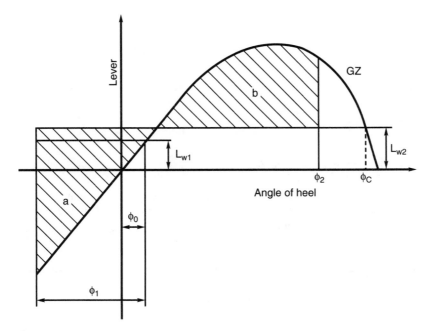

Figure A.4 Weather criterion

Z is the vertical distance from the centre of A to the centre of the underwater lateral area or approximately to a point at $T/2$ in [m].
Δ is the displacement in [t].

The angle ϕ_1 [deg.] is calculated as

$$\phi_1 = 109 \cdot k \cdot X_1 \cdot X_2 \cdot \sqrt{rs}$$

k factor as follows:
  $k = 1.0$ for a round-bilged ship having no bilge or bar keels
  $k = 0.7$ for a ship having sharp bilges
  k according to Table A.1 for a ship having bilge keels, a bar keel or both. A_k is the total overall area of bilge keels, or area of the lateral projection of the bar keel, or sum of these areas [m^2].
X_1 factor as shown in Table A.2.

Table A.1 Factor k

$(A_k \cdot 100)/(L \cdot B)$	0	1.0	1.5	2.0	2.5	3.0	3.5	$\geq$4.0
k	1.0	0.98	0.95	0.88	0.79	0.74	0.72	0.70

Table A.2 Factor X_1

B/T	$\leq$2.4	2.5	2.6	2.7	2.8	2.9	3.0	3.1	3.2	3.4	$\geq$3.5
X_1	1.0	0.98	0.96	0.95	0.93	0.91	0.90	0.88	0.86	0.82	0.80

Table A.3 Factor X_2

C_B	$\leq$0.45	0.50	0.55	0.60	0.65	$\geq$0.70
X_2	0.75	0.82	0.89	0.95	0.97	1.0

$r = 0.73 \pm 0.6\overline{OG}/T$.
$\overline{OG}$ is the distance between the centre of gravity and the waterline [m] ($+$ if the centre of gravity is above the waterline, $-$ if it is below).
s factor as shown in Table A.4.
The rolling period T_r is given by $T_r = 2 \cdot C \cdot B/\sqrt{GM}$; $C = 0.373 + 0.023(B/T) - 0.00043 \cdot L$.

X_2 factor as shown in Table A.3.
  $r = 0.73 \pm 0.6\overline{OG}/T$
  $\overline{OG}$ is the distance between the centre of gravity and the waterline [m] ($+$ if the centre of gravity is above the waterline, $-$ if it is below).
s factor as shown in Table A.4
  The rolling period T_r is given by $T_r = 2 \cdot C \cdot B/\sqrt{GM}$; $C = 0.373 + 0.023(B/T) - 0.00043 \cdot L$.
Intermediate values in Tables A.1 to A.4 should be linearly interpolated.

Table A.4 Factor s

T_r	$\leq$6	7	8	12	14	16	18	$\geq$20
s	0.100	0.098	0.093	0.065	0.053	0.044	0.038	0.035

- For ships operating in areas where ice accretion is likely, icing allowances should be included in the stability calculations. This concerns particularly cargo ships carrying timber deck cargoes, fishing vessels and dynamically supported crafts.

SOLAS (1974)

The damaged stability characteristics of ships are largely defined in the SOLAS Convention (Safety of Life at Sea) (IMO, 1997). Damaged stability is required for nearly all seagoing ships, either on a deterministic or probabilistic basis. The probabilistic approach requires a subdivision index 'A' to be greater than a required minimum value 'R'. 'A' is the total probability of the ship surviving all damages. $A = \Sigma p_i \cdot s_i$, where p_i is the probability that a certain combination of subdivisions is damaged and s_i is the survivability factor ranging from 0 (no survival) to 1 (survival). In 99% of all damage cases of actual designs, s is either 1 or 0 (Björkman, 1995). Sonnenschein and Yang (1993) point out some weaknesses in the SOLAS rules in comparison to U.S. Coast Guard rules. Further discussions of the SOLAS rules, sometimes with examples, are found in Abicht (1988, 1989, 1992) and Gilbert and Card (1990). All ships transporting bulk grain are subject to regulations as documented in Chapter VI of SOLAS (1974), amended in 1994.

MARPOL 73/78

Rule 25 of the MARPOL convention (IMO, 1992) imposes special requirements concerning damage stability for tankers. These requirements are, like some of the SOLAS requirements, probabilistic, but differ in detail; e.g. MARPOL assumes that damage location is as probable everywhere along the ship's length, while SOLAS assumes that damage is more likely in the foreship (Björkman, 1995).

National regulations (Germany)

National regulations usually follow the above international regulations, but may impose additional requirements. German rules are given here as an example.

SBG regulations

In 1984 the SBG (Seeberufsgenossenschaft = German Mariners' Association) issued new regulations for intact stability which consider ship type and cargo type (SBG, 1984). These recommendations refer to the righting arm curve. Table A.5 gives the minimum required values.

- Ships with $L \leq 100\,\mathrm{m}$ and $50° < \phi_0 < 60°$: $h_{30°} = 0.2 + (60° - \phi_0) \cdot 0.01$.
- Cargo-carrying pontoons: $\phi_0 \geq 30°$; $e_{0,\phi_{\max}} \geq 0.07\,\mathrm{m\cdot rad}$.
- Containers as deck cargo: $\overline{GM}' \geq 0.30\,\mathrm{m}$ for $L \leq 100\,\mathrm{m}$, $\overline{GM}' \geq 0.40\,\mathrm{m}$ for $L > 120\,\mathrm{m}$, linear interpolation in between.
- Timber as deck cargo, densely stowed: $\overline{GM}' \geq 0.15\,\mathrm{m}$; $h_{30°} \geq 0.10\,\mathrm{m}$ for $F'/B \leq 0.1$, $h_{30°} \geq 0.20\,\mathrm{m}$ for $F'/B \geq 0.2$, linear interpolation in between. F' is an ideal freeboard, the difference between ideal draught and available mean draught.

Table A.5 Stability requirements of the SBG for cargo ships (summary)

	$h_{30°}$ [m]	$\overline{GM}'$ [m]	$e_{0,30°}$ [m·rad]	$e_{30,40°}$ [m·rad]	$e_{0,40°}$ [m·rad]	ϕ_0 [deg]
General, $L \leq 100$ m	0.20	0.15	0.055	0.03	0.09	50–60
General, 100 m $< L < 200$ m	$0.002L$	0.15	0.055	0.03	0.09	50–60
General, $L > 200$ m	0.40	0.15	0.055	0.03	0.09	50–60
Tugs	0.30	0.60	0.055	0.03	0.09	60

$h_{30°}$	Righting lever at 30° heel
$\overline{GM}'$	Metacentric height corrected for free surfaces
$e_{0,30°}$	Area under static stability curve to 30°
$e_{30,40°}$	Area under static stability curve between 30° and 40°
$e_{0,40°}$	Area under static stability curve to 40°
ϕ_0	Stability range; heeling angle at which righting lever becomes zero again

- Timber as deck cargo, packaged timber: $\overline{GM}' \geq 0.15$ m; $h_{30°} \geq 0.15$ m.
- Coke as deck cargo: $h_{30°}$ is to be increased by 0.05 m.
- Passenger ships:
  Maximum heel angles are:
  10° resulting from passengers crowding to one side
  12° resulting from passengers crowding to one side and turning
  12° resulting from lateral wind pressure.
  The minimum residual freeboard to the bulkhead deck or openable windows must be 0.20 m when the ship is heeled by the above moments. Ships of over 12 m width must show that the lower edges of the windows above the bulkhead deck are not submerged under dynamic wind conditions.
  The heeling moment due to passengers crowding on one side assumes 4 persons/m² for open spaces, otherwise the 'most realistic' assumptions, and 750 N per person plus 250 N luggage (50 N for day trips), centre of gravity 1 m above the deck at the side at $L/2$.
  The heeling moment due to turning is as given for the IMO code of intact stability above.
- Ships with large wind lateral area, except passenger ships:
  The heel angle under side wind is to be calculated.

$$M_{Kr} = p \cdot A \cdot \left(l_w + \frac{T}{2} \right)$$

$p = 0.3$ kN/m² for coastal operation (Bft 9)
$p = 0.6$ kN/m² for short-distance operation (Bft 10)
$p = 1.0$ kN/m² for middle- and long-distance operation (Bft 12)
The heel angle may not exceed 18°. The minimum residual freeboard under heel is 10% of the freeboard for the upright ship.

Further regulations concern tankers, hopper dredgers, ships with self-bailing cockpit or without hatch covers, offshore supply vessels, and heavy cargo-handling.

German Navy stability review

All ships (except submarines) in the German navy are subject to a 'stability review' in which the lever arm curves of righting and heeling moments are compared for smooth water conditions and in heavy seas (Vogt, 1988). The calculation of stability in heavy seas assumes waves of ship's length moving at the same speed and in the same direction as the ship. Seen from the ship, this gives the impression of a standing wave. Different heeling moments and stability requirements—e.g. relating to the inclination achieved—are specified for the following sea conditions:

1. Ship in calm water.
2. Ship on wave crest.
3. Effectiveness of a lever arm curve determined as the mean value from wave crest and wave trough conditions.

Various load conditions form the basis for all three cases. The navy adopted this method of comparing heeling and righting lever arms on the advice of Wendel (1965) who initiated this approach. The stability review can also be used to improve the safety of cargo ships, although it cannot account for dynamic effects. The approach is especially useful for ships with broad, shallow sterns.

References

ABICHT, W. (1988). Leckstabilität und Unterteilung. *Handbuch der Werften* **XIX**, Hansa, p. 13

ABICHT, W. (1989). New formulas for calculating the probability of compartment flooding in the case of side damage. *Schiffstechnik* **36**, p. 51

ABICHT, W. (1992). Unterteilung und Leckstabilität von Frachtschiffen. *Handbuch der Werften* **XXI**, Hansa, p. 281

BJÖRKMAN, A. (1995). On probabilistic damage stability. *Naval Architect*, p. E484

GILBERT, R. R. and CARD, J. C. (1990). The new international standard for subdivision and damage stability of dry cargo ships. *Marine Technology* **27/2**, p. 117

IMO (1992). *MARPOL 73/78*—Consolidated Edition 1991. International Maritime Organization, London

IMO (1995). *Code on intact stability for all types of ships covered by IMO instruments*. International Maritime Organization, London

IMO (1997). *SOLAS*—Consolidated Edition 1997 International Maritime Organization, London

RAHOLA, I. (1939). The Judging of the Stability of Ships. Ph.D. thesis, Helsinki

SBG (1984). Bekanntmachung über die Anwendung der Stabilitätsvorschriften für Frachtschiffe, Fahrgastschiffe und Sonderfahrzeuge vom 24. Oktober 1984. Seeberufsgenossenschaft, Hamburg

SEEFISCH, F. (1965). Stabilitätsbeurteilung in der Praxis. *Jahrbuch Schiffbautechn. Gesellschaft*, Springer, p. 578

SONNENSCHEIN, R. J. and YANG, C. C. (1993). One-compartment damage survivability versus 1992 IMO probabilistic damage criteria for dry cargo ships. *Marine Technology* **30/1**, p. 3

VOGT, K. (1988). Stabilitätsvorschriften für Schiffe/Boote der Bundeswehr. *Handbuch der Werften* **XIX**, Hansa, p. 91

WENDEL, K. (1965). Bemessung und Überwachung der Stabilität. *Jahrbuch Schiffbautechn. Gesellschaft*, Springer, p. 609

Nomenclature

Symbol	Title	Recommended measuring unit
A	Area in general	m^2
A	Rise of floor	m
A_{BT}	Area of transverse cross-section of a bulbous bow	m^2
A_E	Expanded blade area of a propeller	m^2
A_L	Lateral-plane area	m^2
A_M	Midship section area	m^2
A_0	Disc area of a propeller: $\pi \cdot D^2/4$	m^2
AP	Aft perpendicular	
b	Height of camber	m
B	Width in general	m
$\overline{BM}$	Height of transverse metacentre (M) above centre of buoyancy (B)	m
BN	Beaufort number	Bft
C	Coefficient in general	
C_A	Correlation allowance	
C_B	Block coefficient: $\nabla/(L \cdot B \cdot T)$	
C_{BD}	Block coefficient based on depth	
C_{BA}	Block coefficient of aftbody	
C_{BF}	Block coefficient of forebody	
C_{DH}	Volumetric deckhouse weight	
C_F	Frictional resistance coefficient	
C_M	Midship section area coefficient: $A_M/(B \cdot T)$	
C_M	Factor taking account of the initial costs of the 'remaining parts' of the propulsion unit	
C_P	Prismatic coefficient: $\nabla/(A_M \cdot L)$	
C_{PA}	Prismatic coefficient of the aftbody	
C_{PF}	Prismatic coefficient of the forebody	
C_s	Reduced thrust loading coefficient	
C_{Th}	Thrust loading coefficient	
CEM	Concept Exploration Model	
CRF	Capital recovery factor	1/yr
C_∇	Volume–length coefficient	
C_{WP}	Waterplane area coefficient: $A_{WL}/(L \cdot B)$	
CWL	Constructed waterline	

d	Cover breadth	m
D	Moulded depth of ship hull	m
D, D_p	Diameter of propeller	m
D_A	Nozzle outside diameter	m
D_A	Depth corrected for superstructures	m
D_I	Nozzle inside diameter	m
e	Dynamic lever as defined by Rahola	m
E	Dynamic stability	Nm, J
F	Freeboard	m
F	Annual operating time	h/yr
F_n	Froude number: $V/\sqrt{g \cdot L}$	
F_o	Upper deck of a deckhouse	m^2
F_u	Actually built over area of a deckhouse	m^2
FP	Forward perpendicular	
G_{DH}	Deckhouse mass	kg
GL	Germanischer Lloyd	
$\overline{GM}, \overline{GM}_0$	Height of metacentre (M) above centre of gravity (G)	m
h	Water depth	m
h	Lever arm	m
h_{db}	Height of double bottom	m
i	Rate of interest	1/yr
i_E	Half-angle of entrance of waterline	°
i_R	Half-angle of run of waterline	°
I_T	Transverse moment of inertia of waterplane	m^4
J	Advance coefficient	
k	Annual payment	MU/yr
k	Form factor addition	MU/yr
K	Individual payment	MU/yr
k_f	Costs of one unit of fuel	MU/t
k_l	Costs of one unit of lubricating oil	MU/t
k_M	Costs of one unit of engine power	MU/kW
k_{st}	Costs of one unit of installed steel	MU/t
K	Correction factor in general	
K_G	Invested capital	MU
K_M	Costs of main engine	MU
K_{PV}	Present value	MU
$\overline{KB}$	Height of centre of buoyancy (B) above keel (K)	m
$\overline{KM}$	Height of transverse metacentre (M) above keel (K)	m
$\overline{KG}_{StR}$	Height of centre of gravity of the steel hull above keel	m
l	Cover length	m
l	Investment life	yr

L	Length in general	m
L'	Wave forming length	m
L_B	Length of bulb	m
L_D	Length of nozzle	m
L_E	Length of entrance	m
L_{os}	Length over surface	m
L_{pp}	Length between perpendiculars	m
L_R	Length of run	m
L_{wl}	Length of waterline	m
lcb	Distance of centre of buoyancy from midship section	m
M_{Kr}	Heeling moment	Nm
MU	Monetary unit	DM, $, etc.
n	Number of decks	
n	Rate of revolution	min^{-1}
NPV	Net present value	MU
P	Parallel middle body	m
P_B	Brake power	kW
P_D	Delivered power	kW
P_E	Effective power	kW
PWF	Present worth factor	
R	Radius in general	m
R_{AA}	Wind resistance	N
R_n	Reynolds number	
R_F	Frictional resistance	N
R_{PV}	Viscous pressure resistance	N
R_R	Residual resistance	N
R_T	Total resistance	N
s	Height of a parabola	mm
s_f	Specific fuel consumption	$g/(kW \cdot h)$
s_l	Specific lubricant consumption	g/kWh
s_v	Forward sheer height	m
s_h	Aft sheer height	m
S	Wetted surface	m^2
t	Thrust deduction fraction: $(T - R_T)/T$	
t	Trim	m
t	Material strength	mm
t_D	Nozzle thrust deduction fraction	
T	Draught in general	m
T	Propeller thrust	N
T_d	Nozzle thrust	N
V	Speed of ship	kn
V_A	Advance speed of a propeller	m/s
∇	Volume in general	m^3
∇	Displacement volume of a ship	m^3
∇_A	Superstructure volume	m^3

∇_b	Volume of beam camber	m^3
∇_D	Hull volume to depth, D	m^3
∇_L	Hatchway volume	m^3
∇_s	Volume of sheer	m^3
∇_{db}	Volume of double bottom	m^3
∇_{DH}	deckshouse volume	m^3
∇_{LR}	Hold volume	m^3
∇_U	Volume below topmost continuous deck	m^3
w	Wake fraction: $(V - V_A)/V$	
w_d	Nozzle wake fraction	
W	Section modulus	m^3
W_{dw}	Deadweight	t
W_{Agg}	Weight of diesel unit	t
W_{Getr}	Weight of gearbox	t
W_l	Cover weight	t
W_M	Weight of propulsion unit	t
W_o	Weight of equipment and outfit	t
W_{Prop}	Weight of propeller	t
W_R	Weight margin	t
W_{St}	Weight of steel hull	t
W_{StAD}	Weight of steel for superstructures and deckhouses	t
W_{StR}	Weight of steel hull w/o superstructures	t
W_{StF}	Weight of engine foundation	t
W_Z	Weight of cylinder boiler	t
WED	Wake equalizing duct	
WL	Waterline	
y,Y	Offset in body plan of half width plan	
Z	Number of propeller blades	
α	Nozzle dihedral angle	°
η_D	Quasi propulsive efficiency: $R_T \cdot V/P_D$	
η_H	Hull efficiency: $(1 - t)/(1 - w)$	
η_o	Propeller efficiency in open water	
η_R	Relative rotative efficiency	
λ	Wavelength	m
ρ	Mass density: m/∇	t/m^3
τ	Load ratio	
Δ	Displacement mass	t
Δ	Difference (mathematical operator)	
ϕ	Angle of inclination, heel angle	°

Index